SpringerBriefs in Earth System Sciences

SpringerBriefs in Earth System Sciences present concise summaries of cutting-edge research and practical applications. The series focuses on interdisciplinary research linking the lithosphere, atmosphere, biosphere, cryosphere, and hydrosphere building the system earth. It publishes peer-reviewed monographs under the editorial supervision of an international advisory board with the aim to publish 8 to 12 weeks after acceptance. Featuring compact volumes of 50 to 125 pages (approx. 20,000—70,000 words), the series covers a range of content from professional to academic such as:

- A timely reports of state-of-the art analytical techniques
- bridges between new research results
- snapshots of hot and/or emerging topics
- literature reviews
- in-depth case studies

Briefs are published as part of Springer's eBook collection, with millions of users worldwide. In addition, Briefs are available for individual print and electronic purchase. Briefs are characterized by fast, global electronic dissemination, standard publishing contracts, easy-to-use manuscript preparation and formatting guidelines, and expedited production schedules.

Both solicited and unsolicited manuscripts are considered for publication in this series.

More information about this series at https://link.springer.com/bookseries/10032

Laurens M. Bouwer · Doris Dransch ·
Roland Ruhnke · Diana Rechid ·
Stephan Frickenhaus · Jens Greinert
Editors

Integrating Data Science and Earth Science

Challenges and Solutions

 Springer

Editors
Laurens M. Bouwer
Climate Service Center Germany
(GERICS)
Helmholtz-Zentrum Hereon
Hamburg, Germany

Roland Ruhnke
Karlsruhe Institute of Technology
Eggenstein-Leopoldshafen, Germany

Stephan Frickenhaus
Alfred Wegener Institute Helmholtz Centre
for Polar and Marine Research
Bremerhaven, Germany

Doris Dransch
Helmholtz Centre Potsdam - GFZ German
Research Centre for Geosciences
Potsdam, Brandenburg, Germany

Diana Rechid
Climate Service Center Germany
(GERICS)
Helmholtz-Zentrum Hereon
Hamburg, Germany

Jens Greinert
GEOMAR Helmholtz Centre for Ocean
Research Kiel
Kiel, Germany

Helmholtz Association

ISSN 2191-589X   ISSN 2191-5903 (electronic)
SpringerBriefs in Earth System Sciences
ISBN 978-3-030-99545-4   ISBN 978-3-030-99546-1 (eBook)
https://doi.org/10.1007/978-3-030-99546-1

This Springer imprint is published by the registered company Springer Nature Switzerland AG
The registered company address is: Gewerbestrasse 11, 6330 Cham, Switzerland

Foreword by Johann-Christoph Freytag

Over the last two decades, the idea of data-intensive computing, sometimes also called Big Data, which then further developed into data science, has become the foundation for many disciplines to perform data-centric research. While Big Data often refers to the ability to transform, to manipulate, and to access large volumes of data in a scalable manner, data science summarizes all activities that evolve around acquiring, cleaning, integrating, transforming, accessing, analyzing, and visualizing data. The latter activities have been part of numerous research activities in various disciplines for many years. To be able to perform these complex tasks on (almost) arbitrary large inputs has opened up new opportunities for many disciplines to perform their research in an unprecedented manner. For example, it has become feasible to digest vast amounts of incoming data from a large number of sensors in (almost) real time to analyze and guide experiments while they are running. Furthermore, machine learning algorithms such as neural networks require a learning phase which in the past took several months. Current hardware and software potentially together with newly developed concepts reduce the learning phase to minutes or hours, thus making this technology feasible and applicable for many disciplines in finding solutions to complex problems. Examples for such newly developed concepts are the MapReduce paradigm for parallel program execution, or workflows for organizing complex executions of individual tasks ("units of work") in a distributed, heterogeneous compute environment.

Still, using these new concepts and systems developed in computer science becomes challenging when using and applying them in any discipline for various reasons. First, there is the need to ensure a (mental) transfer of often non-trivial concepts and systems into a discipline where they are the means to perform research, rather than being the research subject itself. Second, building complex hardware and software systems obviously goes beyond the skills of applying and programming these for individual applications. They still pose a challenge and require special skills especially when those systems must later deliver results in a reliable and robust manner, and therefore go beyond research prototypes. Third, there is often much effort needed to adapt and/or to extend the computer science concepts and systems in their general form to the specific needs and requirements of a particular application domain.

Already in the book *The Fourth Paradigm—Data-Intensive Scientific Discovery*, edited by Hey, Tansley, and Toll in 2009, the first chapter of the book presents Jim Gray's vision on a new paradigm for performing science in all areas in the future. There, he outlines the clear path of how the progress in hardware and software together with new concepts will transform other sciences through data-intensive computation dramatically, using the term eScience. His vision has become reality in many disciplines, including Earth sciences in particular.

This book is an excellent witness of how researchers in Earth sciences implement Gray's vision to accelerate the generation of new knowledge and insights about our planet. Since the book reflects the results of a longer-term research project, it becomes clear that the usage of computer science concepts and systems is not a straightforward step. Using the new concepts needs clear understanding of domain experts of what is required as a first step, to advance the state of the art in their field. However, the authors master these challenges and clearly show that they understand the newly developed computer science concepts and their use in Earth sciences. In particular, they provide a holistic approach to combine already existing approaches and systems in the different subfields of Earth sciences by the concept of (scientific) workflows. The systematic approach and the detailed discussions and evaluations provided in the different chapters clearly demonstrate the advantage of a holistic approach to study system Earth.

Altogether, the book carefully presents the problems and the chosen solutions, while providing an evaluation and self-assessment of the different phases during the project. The careful reader will identify many challenges and problems left open, which I consider as a strength rather than a weakness of the book. Therefore, one should understand the book as the starting point for further studying system Earth as a highly complex system, with several subsystems connected by many dependencies using data science approaches. Only a holistic approach will advance the understanding of system Earth, for example when developing solutions for such complex problems as the currently discussed climate change.

Enjoy reading this book with its many facets originating in Earth sciences and computer science.

Johann-Christoph Freytag
Department of Computer Science
Humboldt-Universität zu Berlin
Berlin, Germany

Einstein Center Digital Future
Berlin, Germany

Foreword by Hans Pfeiffenberger

Intuitively, the term "Digital Earth" should appeal to contemporary scientists and, even more so, to computer and information technology experts. The consortium from Helmholtz Association's Earth and Environment research area, which organized and wrote this book, as well as other groups around the world striving to implement concepts of Digital Earth, pursues this work mostly based on short-term funding and thus, near-term objectives and deliverables. Research projects such as Digital Earth can contribute substantially to the development of methods and algorithms. When we envision, as a next step, Digital Earths as contributing to societal needs in a comprehensive and sustained way, these projects' resources would be too limited.

If the ultimate objective is not restricted to Digital Germany or other "digital territory," one can identify important societal targets among the UN's Sustainable Development Goals (SDGs). Digital Earth facets might serve, for example, to provide evidence about: climate, biodiversity, health, water and food supply, mineral resources, etc. Major organizations such as the Helmholtz Association are provided with the funding and charged with a mission and the multidisciplinary approaches necessary for progress on finding solutions to these complex issues.

Occasionally, those building Digital Earths and those applying them to socially relevant issues should take the time and philosophize. They should ask themselves: what do our digitization concepts and their implementation aim at, what can they achieve, what are their inherent limitations, or (most important but often most difficult) what dangers might arise?

For millennia, individuals, countries, and institutions developed more precise, more useful representations of the world surrounding them. Produced from perspectives at the time, these representations would be expressed and shared (or, not) as written text or as ever-more sophisticated maps.

"Gallia est omnis divisa in partes tres, quarum unam incolunt Belgae, aliam Aquitani, tertiam qui ipsorum lingua … ". This famous first sentence of Julius Caesar's "De Bello Gallico" emphasizes the importance of knowing the world. Rome, in its role as regional superpower, required up-to-date information about physical distances (and obstacles), populations, customs, and languages to maintain military dominance.

Maps from the sixteenth century and seventeenth century CE demonstrate the utility of continuously extended and corrected maps to support commercially motivated travel on land or navigation at sea. Such graphic and narrative guides, produced by and for competing countries, were often kept as military and economic secrets.

Digital transformations already deeply affect our perception of our world, even into everyday life. We scientists must recognize that, beyond our traditional roles as sources of research-quality texts and maps, governments and corporations rapidly add and exploit big datasets and models, computer-based visualization, data analytics, and machine learning and Artificial Intelligence tools that influence and shape individual and national perspectives of our world.

We can and must learn from historical precedents that contemporary worldviews were shaped by particular perspectives and interests of those creating them. Caesar and Roman predecessors established and promulgated an entity called Gallia, inhabited by three major groups of people. Would people inhabiting Gallia have assigned themselves to one of those three groups? Would they have recognized or named a geographic entity of Gallia? Was the concept of Gallia of any relevance to them?

Likewise, maps in use before the modern era of commercial globalization feature Europe or the Atlantic Ocean at the center. In particular, using Mercator projections for anything other than navigation can create biased geographic (areal) conclusions. Limitations of these well-known projection quirks remain obvious to many but underappreciated by most.

Which analogous traps will emerge and persist because of new Digital Earths?

Will terminologies and ontologies we create to describe and categorize data carry and promote "western" views of the world and its interests? For example, will we produce global remote-sensing-based maps using lists of large-scale globally traded crops, while neglecting traditional small-holder crops which might be better at nourishing self-sustaining communities? Will we perpetuate misleading terms such as "rare diseases" which, too often, indicate economically unimportant diseases? Will we miss important opportunities to investigate non-pandemic diseases or to track their zoonotic vectors? Will visualizations, dependent on interpolations and extrapolations of globally patchy data, mislead or support misconceptions about under-served parts of the world?

Many countries, for political, economic, or "status" reasons, withhold or manipulate national data or delay delivery beyond a useful timeline. Patchy, delayed and (by incompetence or worse) manipulated data will constitute some of the basis of SDG-oriented Digital Earths. All data from all sources must be vetted for completeness, accuracy, and trustworthiness and, to facilitate this level of scrutiny, must be as openly accessible as technically and ethically feasible.

Organizations, including the Helmholtz Association, setting out to create or co-create Digital Earths with far-reaching objectives must be prepared to invest heavily in careful data collection, responsible development of algorithms and code, and enhanced quality assurance. While these operations rarely qualify as research, and thus rarely provide appropriate academic recognition and reward to skilled providers, recent examples such as the consortium of researchers associated with the Global

Carbon Project, producing much-lauded and fully citable annual global greenhouse gas budgets, provide encouraging examples.

Digital Earth developers will contribute to identifying essential products and globally agreed-upon "essential variables," particularly those relevant to SDGs. They need to engage constructively with collaborators around the globe to stimulate viable long-term capabilities and capacities to both feed and use advanced concepts, digital products, and technologies as described in this book.

Hans Pfeiffenberger
Independent Consultant
Bremen, Germany

Acknowledgments

The Digital Earth project has been an exciting and very valuable journey, to create a step-change in connecting data science and Earth system sciences. We had unique and intense collaboration between eight centers of the Helmholtz Association in the field of Earth and Environment. Our success would not have been possible without the support and dedication of several individuals.

We say thanks to the many partners and collaborators in the project consortium, that have developed innovative workflows, visualizations, and diverse applications of data science methods, that are reported in this book. We thank our colleague Dr. Daniela Henkel for her great support as project manager. The Helmholtz Association is thanked for financial support of the Digital Earth project through the Impulse and Networking Fund, funding code ZT-0025. We also thank Springer for supporting our book production.

Finally, we especially thank Prof. Johann-Christoph Freytag and Dr. Hans Pfeiffenberger for their time to critically and constructively review our book, and for making several suggestions, which has helped to substantially improve the structure and presentation of our findings.

Hamburg, December 2021

Laurens M. Bouwer
Doris Dransch
Roland Ruhnke
Diana Rechid
Stephan Frickenhaus
Jens Greinert

Contents

1   **Data Science and Earth System Science**   1
Wolfgang zu Castell, Roland Ruhnke, Laurens M. Bouwer,
Holger Brix, Peter Dietrich, Doris Dransch, Stephan Frickenhaus,
Jens Greinert, and Andreas Petzold

2   **The Digital Earth Project: Focus and Agenda**   7
Roland Ruhnke, Diana Rechid, Doris Dransch,
Laurens M. Bouwer, Holger Brix, Peter Dietrich,
Stephan Frickenhaus, Jens Greinert, Daniela Henkel,
Andreas Petzold, and Wolfgang zu Castell

3   **Data Analysis and Exploration with Visual Approaches** 17
Everardo González Ávalos, Doris Dransch, Nicola Abraham,
Valentin Buck, Daniel Eggert, Tom Kwasnitschka, Daniela Rabe,
Flemming Stäbler, and Viktoria Wichert

4   **Data Analysis and Exploration with Computational Approaches** 29
Viktoria Wichert, Laurens M. Bouwer, Nicola Abraham,
Holger Brix, Ulrich Callies, Everardo González Ávalos,
Lennart Christopher Marien, Volker Matthias, Patrick Michaelis,
Daniela Rabe, Diana Rechid, Roland Ruhnke, Christian Scharun,
Mahyar Valizadeh, Andrey Vlasenko, and Wolfgang zu Castell

5   **Data Analysis and Exploration with Scientific Workflows** 55
Doris Dransch, Daniel Eggert, Nicola Abraham,
Laurens M. Bouwer, Holger Brix, Ulrich Callies,
Thomas Kalbacher, Stefan Lüdtke, Bruno Merz, Christine Nam,
Erik Nixdorf, Daniela Rabe, Diana Rechid, Kai Schröter,
Bente Tiedje, Dadiyorto Wendi, and Viktoria Wichert

6   **The Digital Earth Smart Monitoring Concept and Tools**   85
    Uta Koedel, Peter Dietrich, Philipp Fischer, Jens Greinert,
    Ulrich Bundke, Ewa Burwicz-Galerne, Antonie Haas,
    Isabel Herrarte, Amir Haroon, Marion Jegen, Thomas Kalbacher,
    Marcel Kennert, Tobias Korf, Ralf Kunkel, Ching Yin Kwok,
    Christoph Mahnke, Erik Nixdorf, Hendrik Paasche,
    Everardo González Ávalos, Andreas Petzold, Susanne Rohs,
    Robert Wagner, and Andreas Walter

7   **Interdisciplinary Collaboration**   121
    Nike Fuchs and Gesche Krause

8   **Evaluating the Success of the Digital Earth Project**   131
    Laurens M. Bouwer, Diana Rechid, Bernadette Fritzsch,
    Daniela Henkel, Thomas Kalbacher, Werner Köckeritz,
    and Roland Ruhnke

9   **Lessons Learned in the Digital Earth Project**   145
    Jens Greinert, Daniela Henkel, Doris Dransch,
    Laurens M. Bouwer, Holger Brix, Peter Dietrich,
    Stephan Frickenhaus, Andreas Petzold, Diana Rechid,
    Roland Ruhnke, and Wolfgang zu Castell

Chapter 1
Data Science and Earth System Science

Wolfgang zu Castell, Roland Ruhnke, Laurens M. Bouwer, Holger Brix,
Peter Dietrich, Doris Dransch, Stephan Frickenhaus, Jens Greinert,
and Andreas Petzold

Abstract Data-driven science has turned into a fourth fundamental paradigm of
performing research. Earth System Science, following a holistic approach in unrav-
eling the complex network of processes and interactions shaping system Earth,
particularly profits from embracing data-driven approaches next to observation and
modeling. At the end, increasing digitalization of Earth sciences will lead to cultural
transformation towards a Digital Earth Culture.

Keywords Data analysis · Data exploration · Earth System Science · Data
science · Digitalization · Machine learning

W. zu Castell (✉) · D. Dransch
Helmholtz Centre Potsdam—GFZ German Research Centre for Geosciences, Potsdam, Germany
e-mail: wolfgang.castell@gfz-potsdam.de

R. Ruhnke
Karlsruhe Institute of Technology, Eggenstein-Leopoldshafen, Germany

L. M. Bouwer
Climate Service Center Germany (GERICS), Helmholtz-Zentrum Hereon, Hamburg, Germany

H. Brix
Helmholtz-Zentrum Hereon, Geesthacht, Germany

P. Dietrich
Helmholtz Centre for Environmental Research—UFZ, Leipzig, Germany

S. Frickenhaus
Helmholtz Centre for Polar and Marine Research, Alfred Wegener Institute, Bremerhaven,
Germany

J. Greinert
GEOMAR Helmholtz Centre for Ocean Research Kiel, Kiel, Germany

A. Petzold
Forschungszentrum Jülich GmbH, Jülich, Germany

W. zu Castell
Helmholtz Zentrum München, German Research Center for Environmental Health, Neuherberg,
Germany

© The Author(s) 2022   1
L. M. Bouwer et al. (eds.), *Integrating Data Science and Earth Science*,
SpringerBriefs in Earth System Sciences,
https://doi.org/10.1007/978-3-030-99546-1_1

1.1   Introduction

When Al Gore coined the term Digital Earth about thirty years ago (Gore 1998), he envisaged a holistic tool for Earth system understanding, exploration and education. Imagining a child visiting an exhibition, he sketched the idea of a comprehensive framework for data integration and analysis, allowing for an overall perspective of planet Earth to dive into and refine with additionally enriched data wherever interest is leading to. As far as maps and imagery are concerned, services such as Google Earth^{TM} became well established in the meantime. Whereas the ability to dive deeper and deeper into details and to explore ever more datasets using this tool, a real Digital Earth is still a vision to be realized.

1.2   Data Science

Integrative, exploratory data analysis has been established as a fourth paradigm of science, next to theory, experiment and simulation (Gray 2009). Indeed, data-driven analysis has led to new insights into several fields of research, in particular, in those fields which by their very nature are lacking a comprehensive underlying theory. Data science "focuses on the processes and systems that enable the extraction of knowledge or insights from data in various forms, either structured or unstructured" (Berman et al., 2016 p. 2). As such, data science utilizes computer science, statistics, machine learning, visualization and human–computer interaction to collect, clean, integrate, analyze and visualize data, as well as to interact with data to create insight into some problem(s) in the real world.

   Data-driven approaches to knowledge discovery have penetrated into almost every field of empirical science. Two major developments have paved the way for this radical transformation: first, through the evolution of the World Wide Web, data sources have become available on an unprecedented scale. Using the Internet, access to data sources has been substantially facilitated with more and more data sources becoming available. At the same time, the parallel development of computing technology allowed the processing of an increasing amount of data, allowing researchers to incorporate more data into their models and ingest huge datasets in an automated way. Both of these prerequisites eventually allowed researchers in artificial intelligence to build models which otherwise would not have been feasible to train due to their large number of parameters. Thus, sufficient computing power and the availability of huge amounts of data enabled a switch of paradigm, leading to models of artificial intelligence, predicting possible patterns of interest without the need of an underlying theory. This is particularly true for deep learning networks, the advancement of which developed closely with massively parallel computing technology reaching a commodity level.

In the end, it seems like Al Gore's vision of Digital Earth is but a fingertip away from becoming reality. However, the complexity of a challenge cannot be assessed without taking the first steps toward the goal.

1.3 Earth System Science

Earth System Science, with its historically subdivided disciplines that are based on the Earth compartments, will significantly benefit from integrative data-driven science. Environmental changes are the result of a complex interaction of natural and anthropogenic processes on a wide variety of temporal and spatial scales. Understanding and quantifying these changes must be based on trustworthy and well-documented observations that capture the entire complexity of the Earth system. This includes the manifold interactions between the atmosphere, land and ocean, including the impacts on all forms of life. Targeted environmental research projects and continuously operating multivariate research infrastructures designed to monitor all components of the Earth system are crucial pillars for environmental scientists in their quest for understanding and interpreting the complex Earth system, together with numerical simulations.

Therefore, data in Earth System Science readily complies with four of the 5 Vs of Big Data: volume, velocity, variety and veracity. Space-based observation systems produce a high volume of data at a speed of change (velocity), which increases with every new mission being started. The variety of geospatial information is relying on specialized infrastructures being capable of honoring the spatio-temporal structure of the data (Schade et al. 2020). Due to the global scale and need for long time series, Earth sciences, in particular climate research, have to deal with uncertainty of data on a regular basis (veracity). However, the fifth V, value, can only be extracted when data is turned into knowledge, helping to answer the pressing questions of society (van Genderen et al. 2020).

Making accurate predictions and providing solutions for current questions related, e.g., to climate change, water, energy, biodiversity, food security and the development of scientifically based mitigation and adaptation strategies in the context of climate change and geo hazards are important requests toward the Earth science community worldwide. In addition to these society-driven questions, Earth System Sciences are still strongly motivated by the eagerness of individuals to understand processes, interrelations and tele-connections within small subsystems, between subsystems and the Earth system as a whole. Understanding and predicting temporal and spatial changes and their inherent uncertainties in the above-mentioned micro- to Earth spanning scales are the key to understanding Earth ecosystems. Reliable, high-quality and high-resolution data across all scales (seconds to millions of years; millimeters to 1,000s of km) has to be utilized in an integrative approach enhancing the ability to integrate data from different disciplines, between Earth compartments, and across interfaces.

1.4 Challenges

While embarking on the adventure toward building Digital Earth, we must not stop at collecting data and providing access to various data sources. Data acquisition needs to resolve issues of metadata standards, referencing datasets as well as providing tools for data conversions and data management. High-quality data also needs to be enriched with information on data acquisition technologies, such as error tolerances of sensors and measuring artifacts. Following an Internet of Things (IoT) paradigm, workflows have to be matured toward SMART monitoring, including anomaly detection methods and spatio-temporal imputation.

Taking into account the substantial role of models in Earth System Sciences, computational challenges follow. Simulations need to be run on a sustainable basis with proper methods of parameter tuning. With computing technology changing at a higher rate, legacy code and model libraries have to be adapted to new computing hardware. Thus, Earth scientists providing highly optimized codes have to work in a co-design manner with computer engineers (Schulthess 2015). Splitting code into a backbone part which is obviously closer to the underlying technology platform and a frontend library including application programming interfaces (APIs) will allow scientists to concentrate on their data analysis tasks. At the same time, application programmers can use descriptive programming languages such as Python, leaving imperative programming to the backend.

In the future, geospatial information infrastructures will have to be adjusted in order to cope with rapid changes in computing technology and at the same time scale with an increasing diversity of applications (Bauer et al. 2021). Closely linking model-based simulation with data-driven analysis and prediction will allow to address questions of increasing complexity as resulting from the incorporation of scientific domains lacking an underlying theoretical foundation. Data-driven approaches may also be used to avoid costly simulation runs on high-end HPC systems or to deal with larger gaps in datasets.

However, within a data-centric approach, dealing with large, distributed datasets by means of programming, is unavoidable. Minimizing data movement in algorithms has to be considered as well as making use of data hierarchies (see Schulthess 2015).

Data alone is not sufficient for gaining new insight and knowledge. Many machine learning methods rely on high-quality, annotated data being available for training. Obtaining high-quality, labeled data typically is a tedious task. In order to scale such tasks to a global and just-in-time level, scientists have to be released from doing repetitive, automatable work. Incremental learning techniques have to be developed, filling gaps in data streams, providing reliable labels, as well as sorting out minor quality measurements. Citizen science projects such as PlanktonID (Christiansen & Kiko, 2016) have proven that getting the public involved, large gains can be obtained in combining machine prediction with human perception. This is just one example showing that successful data science approaches reach beyond classical data analysis. At the end, it is the way we interact with data, which will push us to the next level. Being visual beings, new approaches for visual data exploration, technology will enable users to explore complex datasets and set off to new exploration journeys. For such technology to be developed, interdisciplinary teams of Earth scientists, AI specialists and visualization experts have to join forces in modeling data exploration workflows and identifying entry points for technological support.

1.5   Digital Earth Culture

Working in cross-domain teams, making use of the diversity of expertise will be a key requirement of realizing Digital Earth. A new culture of scientific cooperation has to be implemented (Dai et al. 2018). From a slightly broader perspective, working toward Digital Earth will become an instantiation of digital transformation in Earth System Sciences. Making use of digitalization in order to release humans from automatable tasks, building on human creativity and supporting new insight by data-driven hypothesis making will transform knowledge extraction in Earth System Sciences.

Co-creative processes and agile cycles will become the new way of pursuing science. Cross-disciplinary cooperation will advance tools for scientific research, and advanced tools will foster creativity in Earth System Sciences. In general, digitalization and open science will cross-fertilize each other. With results of scientific work being shared, scientific progress will be fostered (Helmholtz Open Science Office, 2021). The complexity of System Earth will never be captured by a single domain perspective alone. To understand the interplay of Earth's compartments and to provide insight into consequences of anthropogenic influence, a combined effort of scientific diversity is needed. At the end, a fully operational digital twin of System Earth might result, seamlessly fusing data from various sources and allowing users to interact with the data, to explore, to learn and to admire the wonder of planet Earth.

References

Bauer P, Dueben PD, Hoefler T, Quintino T, Schulthess TC, Wedi NP (2021) The digital revolution of Earth-system science. Nat Comput Sci 1:104–113. https://doi.org/10.1038/s43588-021-000 23-0

Berman F, Rutenbar R, Christensen H, Davidson S, Estrin D, Franklin M, Hailpern B, Martonosi M, Raghavan P, Stodden W, Szalay A (2016) Realizing the potential of data science. Final Report from the National Science Foundation Computer and Information Science and Engineering Advisory Committee Data Science Working Group. National Science Foundation. https://www.nsf.gov/cise/ac-data-science-report/CISEACDataScienceReport1.19.17.pdf. Accessed 1 October 2021

Christiansen S, Kiko R (2016) PlanktonID. https://planktonid.geomar.de

Dai Q, Shin E, Smith C (2018) Open and inclusive collaboration in science: a framework. OECD Science, Technology and Industry Working Papers, No. 2018/07. OECD Publishing, Paris. https://doi.org/10.1787/2dbff737-en

Gore A (1998) The Digital Earth: understanding our planet in the 21st century. Speech given at the California Science Center, Los Angeles, California, on January 31, ESRI. http://portal.openge ospatial.org/files/?artifact_id=6210. Accessed 1 October 2021

Gray J (2009) Jim Gray on eScience: a transformed scientific method. In: Hey T, Tansley S, Tolle K (eds) The fourth paradigm: data-intensive scientific discovery. Published by Microsoft Research, October 2009, ISBN: 978-0-9825442-0-4. https://www.microsoft.com/en-us/research/ publication/fourth-paradigm-data-intensive-scientific-discovery/

Helmholtz Open Science Office (2021) Open Science und Digitaler Wandel gehen Hand in Hand (German). Potsdam. https://gfzpublic.gfz-potsdam.de/pubman/item/item_5008705

Schade S, Granell C, Vancauwenberghe G, Keßler C, Vandenbroucke D, Masser I, Gould M (2020) Geospatial information infrastructures. In: Guo H et al (eds) Manual of Digital Earth. Springer, Singapore. https://doi.org/10.1007/978-981-32-9915-3_5

Schulthess TC (2015) Programming revisited. Nature Phys 11:369–373. https://doi.org/10.1038/ nphys3294

van Genderen J, Goodchild MF, Guo H, Yang C, Nativi S, Wang L, Wang C (2020) Digital Earth challenges and future trends. In: Guo H et al (eds) Manual of Digital Earth. Springer, Singapore. https://doi.org/10.1007/978-981-32-9915-3_26

Chapter 2
The Digital Earth Project: Focus and Agenda

Roland Ruhnke, Diana Rechid, Doris Dransch, Laurens M. Bouwer, Holger Brix, Peter Dietrich, Stephan Frickenhaus, Jens Greinert, Daniela Henkel, Andreas Petzold, and Wolfgang zu Castell

Abstract Digital Earth is a project funded by the German Helmholtz Association with all centers of the research field Earth and Environment involved. The main goal of the Digital Earth project is to develop and bundle data science methods in extendable and maintainable scientific workflows that enable natural scientists in collaboration with data scientists to achieve a deeper understanding of the Earth system. This has been achieved by developing solutions for data analysis and exploration with visual and computational approaches with data obtained in a SMART monitoring approach and modeling studies, accompanied by a continuous evaluation of the collaboration processes. In this chapter, the history, setup, and focus of the Digital Earth project are described.

R. Ruhnke (✉)
Karlsruhe Institute of Technology, Eggenstein-Leopoldshafen, Germany
e-mail: roland.ruhnke@kit.edu

D. Rechid · L. M. Bouwer
Climate Service Center Germany (GERICS), Helmholtz-Zentrum Hereon, Hamburg, Germany

D. Dransch
Helmholtz Centre Potsdam—GFZ German Research Centre for Geosciences, Potsdam, Germany

H. Brix
Helmholtz-Zentrum Hereon, Geesthacht, Germany

P. Dietrich
Helmholtz Centre for Environmental Research—UFZ, Leipzig, Germany

S. Frickenhaus
Helmholtz Centre for Polar and Marine Research, Alfred Wegener Institute, Bremerhaven, Germany

J. Greinert · D. Henkel
GEOMAR Helmholtz Centre for Ocean Research Kiel, Kiel, Germany

A. Petzold
Forschungszentrum Jülich GmbH, Jülich, Germany

W. zu Castell
Helmholtz Zentrum München, German Research Center for Environmental Health, Neuherberg, Germany

© The Author(s) 2022
L. M. Bouwer et al. (eds.), *Integrating Data Science and Earth Science*,
SpringerBriefs in Earth System Sciences,
https://doi.org/10.1007/978-3-030-99546-1_2

Keywords Data analysis · Data exploration · Monitoring · Collaboration · Interdisciplinary · Workflow · Earth science · Data science · Digitalization

2.1 History of the Project

The Digital Earth project was initiated by the Helmholtz Association (see Box 2.1) in advance of the joint research program 2021–2028, to take some of the ideas and challenges described in Chapter 1 regarding data science, digitalization, and Earth System Science. The vision of the project was to foster interdisciplinary collaboration and to identify and adapt in a strongly interrelated approach methods, workflows, and applications that are true "game-changers" for studying the Earth system.

To achieve this, the eight Helmholtz Centers in the research field Earth and Environment (Alfred Wegener Institute Helmholtz Centre for Polar and Marine Research Bremerhaven (AWI), Forschungszentrum Jülich (FZJ), GEOMAR Helmholtz Centre for Ocean Research Kiel, Helmholtz Centre Potsdam German Research Centre for Geosciences (GFZ), Helmholtz-Zentrum Hereon, Karlsruhe Institute of Technology (KIT), Helmholtz-Zentrum München German Research Center for Environmental Health (HMGU), and Helmholtz Centre for Environmental Research (UFZ)) were asked in 2016 to develop a joint proposal "Digital Earth" as part of a call on future research topics within the Helmholtz Association Initiative and Networking Fund.

Already during the definition phase, it became obvious that focusing on a common direction for a joint proposal was challenging: the involved scientists of each center had different disciplinary backgrounds, expectations, and views on such a project. While some were able to contribute precise geoscientific research questions related to observation or model data from their discipline and their institutional background, others were interested in contributing methods of data science or software engineering for data exploration, and again, others were contributing a perspective on developing data infrastructures for data-intensive science within the project. Consequently, the heterogeneous interests and possible engagements were not fully harmonized toward an effective proposal development, but they rather had to be integrated into a common development and a common understanding of the project goals. To achieve this, a cross-disciplinary and cross-compartment approach was identified as useful, allowing for a holistic view on the coupled compartments of the Earth system. The strength of such an approach is the diversity of perspectives resulting in the challenge to create frameworks for fruitful long-term collaboration within the project.

Box 2.1: The Helmholtz Association at a Glance

- Germany's largest research organization.
- Named after Hermann von Helmholtz (1821–1894), one of the last great scientific generalists.

- Annual budget of ~ € 4.5 billion.
- ~ 39,000 employees.
- World-class science infrastructure.
- 18 independent research centers all over Germany.
- Research plays a key role in identifying reliable answers that benefit society, science, and the economy.
- Six fields of research focus on the major societal challenges of our time—such as the digital revolution, climate change, energy transition, transport in the future, and the battle against severe and widespread diseases and work on developing sustainable solutions for the future. In doing so, Helmholtz covers the entire spectrum from basic to application-oriented research while applying an interdisciplinary approach.
- The Helmholtz Association cooperates with leading research institutions at the national and international level and is committed to the highest standards of talent management at all levels and the promotion of early-career researchers.
- New knowledge can only benefit society and the economy if it is transferred and therefore made usable. For this reason, transferring knowledge and technology and promoting innovation are of extraordinary importance to us.
- 400 new patents are filed every year.
- Approximately 20 new high-tech spin-offs per year.

With the knowledge of the aspects of digitalization described in Chapter 1 and especially the challenges associated with it in mind, the discussion of redefining the aims in favor of reusable, framework-based concepts, the potential of artificial intelligence and advanced visualization methods would play an important role in the Digital Earth project. In addition, it appeared natural to manifest the goal of reusability and sustainability to support the networking process and long-term collaboration explicitly in a dedicated task, since the long-term perspective of further collaboration within the joint research program 2021–2027 was given immanently in the research field Earth and Environment (see Box 2.2).

Box 2.2: Helmholtz Research Program "Changing Earth – Sustaining Our Future"
Climate change, the extinction of species, environmental pollution, and the increasing vulnerability of a technological society to natural disasters are among the greatest challenges of our time. We take a systemic approach to researching our natural environment—from the land surface and the oceans to the most remote polar regions. After all, it will only be possible to plot a course

into a sustainable future with in-depth knowledge of the Earth system, innovative technologies, strategic solutions, and evidence-based recommendations for policymakers.

Seven Helmholtz Centers are collaborating to gather deep insights into the complex relationships between the processes that take place on our planet. What are the causes and effects of global environmental changes? How can natural resources be used sustainably? How can we protect ourselves more effectively from disasters and natural hazards like droughts, heavy rainfall, storms, floods, and earthquakes? We aim to develop solutions and strategies to help humankind adapt to changing environmental conditions, to minimize global threats like climate change, and to understand the potential impact of these risks—not only for the environment but also for the economy and society.

2.2   Focus of Digital Earth

The Digital Earth project addresses the challenge of digital transformation in Earth science. The central goal of the project is to enable Earth scientists to (a) develop methods to link data across compartmental boundaries across spatial and temporal scales; (b) establish coherent data flows and analysis workflows; and (c) develop approaches to guide data acquisition in the field by linking various field and model data. The central question of the project is: How can data science contribute to the goals and improve scientific results? This is the fundamental question asked by the natural scientists toward data science.

Therefore, the Digital Earth project is not directed to develop entirely new data science methods and technologies such as new machine learning algorithms or visualization techniques. The innovative aspect is to link natural science and data science and to develop cross-boundary approaches focusing on three main areas, as they are essential: (i) data analysis and exploration; (ii) data collection and monitoring; and (iii) collaborative interdisciplinary working, which is of special importance for the digital transformation.

Developing, advancing, and adopting means that enable this vision are the tasks of Digital Earth to transfer knowledge and close gaps between the two disciplines of Earth science and data science. Here, two scientific ambitions merge, one from the Earth science community that wants to have data science approaches available for their investigations, and the other ambition from data scientists that wants to advance data science methods in itself, but also to make them more easily adaptable to specific scientific requirements. A dialogue is required, and a long-term and sustainable cooperation and problem-solving culture need to be established, in which questions can be put forward and iteratively worked on, solutions get tested and finally adopted. This approach, which is tailored to the needs of the Earth scientists, includes faster

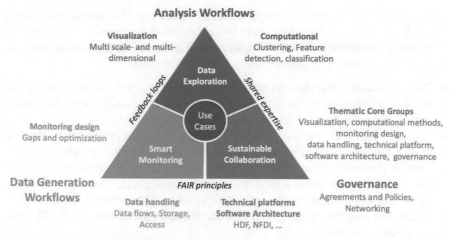

Fig. 2.1 Three cornerstones of the Digital Earth project

and easier-to-use applications, the development/promotion of best methodologies, the adjustment and extension of existing applications, and the implementation of automatization.

Within the focus triangle of Digital Earth (Fig. 2.1), we address several issues: the reuse of data and methods/tools by a broad scientific community reaching far beyond the researchers directly involved in the generation of data and methods/tools, FAIR principles (Findable, Accessible, Interoperable, and Re-usable), quality assessment, visual and computational data exploration, interpolation and integration of data from in situ measures and simulation models, and scientific workflows.

Within several showcases, we adapted and enhanced several data science methods to address challenges we have to face in the investigation of the System Earth. We address challenges related to the following three topics:

- Data analysis and exploration;
- Data collection and monitoring;
- Collaborative interdisciplinary working.

These three topics and their challenges are discussed in the next sections, and the separate chapters in this book that are dedicated to solutions developed in Digital Earth for addressing these challenges are introduced.

2.2.1   Data Analysis and Exploration

Challenges for data analysis and exploration addressed with **visual approaches** (see Chapter 3): The incessant processes shaping our Earth's environment are determined by an interplay of diverse phenomena of physical, chemical, and biological nature, with ranges of action that span from planetary scales into the microscopic realm. As such, the study of the geoscientific processes and the interplay of determining phenomena rely on the analysis of highly diverse kinds of data from many sources. The following challenges arise for analysis and exploration:

- **The need to establish connections between causes and consequences of geoscientific processes**: The connections become evident only when observations from various disciplines and sources are brought into the relationship.
- **The need to retain a sense of spatial and temporal coherence across different scales**: The sense of spatial and temporal coherence is easily lost, when simultaneously regarding information at different scales such as sediment samples that encode information at cm scale, while remote sensing data does so in km scale. In the temporal dimension, an underwater sediment plume can arise and settle in a matter of minutes, while global climatic phenomena are compared with each other across decades.
- **The need for suitable means to integrate a variety of heterogeneous spatio-temporal datasets**: Scientists have to be supported in creating a "holistic view" on processes and related phenomena.

Digital Earth addresses these challenges with visualization. A main advantage of visualization is its ability to parallel display data even if the data is heterogeneous in scale, variables or accuracy. We applied various visualization techniques and environments and adopted them to our geoscientific requirements. The outcomes are tools for interactive data exploration based on (a) multiple linked view techniques; (b) web-based technologies for real-time exploration of data across spatial and temporal scales; and (c) immersive visualization.

Challenges for data analysis and exploration addressed with computational approaches (see Chapter 4): artificial intelligence and machine learning methods are increasingly applied in Earth system research, for improving data analysis, and model performance, and eventually system understanding. Digital Earth focuses on:

- **The need to extract relevant information/features using machine learning**: For various observational features, no labeled data collections exist. Such labels are, however, important to classify specific observations using prior knowledge, for instance. Using sparse datasets and machine learning methods, alternative ways were found to broaden data availability and derive new, crucial information from existing data. We used examples to map river levees in Germany for which no consistent data was available, and for locating ammunition on the sea-bed.
- **The need to approximate complex processes with machine learning**: Some processes in the Earth system are too complex or computationally costly for large-scale or multiple simulations in models. Here, machine learning alternatives can replace some of these (partly unknown) processes. We present applications for atmospheric methane and ethane concentrations through a neural network and for combining highly heterogeneous data to simulate relations between extreme temperatures and health outcomes.

- **The need of point-to-space extrapolation**: For many applications in Earth systems research, the extrapolation of point to space and local measurements to regional or global fluxes are essential. We employ different computational approaches, for analysis and processing of point observations of methane emissions in order to be comparable with global atmospheric emissions as observed/estimated in global databases, and the functionalities of advanced approaches for point-to-space extrapolation.
- **The need of anomaly and event detection across heterogeneous datasets**: Events and anomalies are important to detect in Earth systems for scientific and practical applications. The huge amount of data and the associated heterogeneity, requires analytic approaches that automate data analysis and still provides relevant results. We present two approaches to detect and understand events in coastal and river waters that are based on this principle: one to assess the similarity of river flood events using multiple atmospheric, hydrologic, and other variables and another that combines observational and model data to detect river plumes at sea at the end of a riverine flood event chain and tracks their spatial and temporal extent.

Challenges for data analysis and exploration addressed with **scientific workflows** (Chapter 5). The challenges include:

- **The need for enhanced work environments that integrate methods and tools into seamless data analysis chains** which allow scientists to comprehensively analyze and explore heterogeneous, distributed datasets. Currently, scientific data analysis is often characterized by performing the analytical tasks in single isolated steps with several isolated tools. This isolated work environment hinders scientists to extensively exploit and analyze the available data.
- **The need for sharing and reuse of analytical methods and tools**: Scientific data analysis and exploration often require specific, highly tailored methods and tools; many of them are developed by geoscientists themselves. Often the methods and tools can hardly be shared and re-used since they lack state-of-the-art computer science methods. The analysis methods and tools are not available for others and have to be invented again and again.
- **The need to exploit data across the various scientific disciplines in Earth System Science**: To answer complex scientific questions, data from various sources has to be integrated, but also the data analysis approaches itself that extract information from the data have to be integrated across disciplines. Integration is necessary on two levels: integration on the technical executable level, but also on the conceptual scientific level.
- **The need to transform science into digital science**: The transformation of science into digital science has been an ongoing process for many years. Suitable means are required to facilitate this transformation and to support collaboration of computer- and geo-experts.

Digital Earth applied the concepts of scientific workflows and component-based software engineering to address these challenges and needs. We adapted and proved the concepts in our geoscientific work environment and assessed how the approaches can tackle the challenges. Within the showcase "cross-disciplinary investigation of flood events," we developed (a) several data analysis workflows on a conceptual and digital level and (b) the component-based Data Analytics Software Framework (DASF). The outcome is the Digital Earth Flood Event Explorer that allows investigating floods from several perspectives and that exemplarily shows how scientific workflows and component-based software engineering can improve scientific data analysis.

2.2.2   Data Collection and Monitoring

Challenges for data collection and monitoring addressed with **SMART monitoring approaches** (Chapter 6). The challenges include:

- **The need for SMART Sensors**: Advancing and developing sensors that have real-time data (pre)processing capacities and are linked in a self-organizing sensor network is still a challenging technological task. Automated event detection, drift correction, and failure detection are possible but still rarely done. Real-time data connections and centralized visualization and analyses are more and more established, but the real challenge is that such SMART sensors and sensor networks become easy to use and the standard way of acquiring multiparameter data in the field.
- **The need for a SMART DataFlow**: An easy to use, scalable and adaptable way of receiving data from sensors and re-distributing them through various channels and means also in real time is the challenge for an efficient SMART Monitoring DataFlow. Standardized and largely automated procedures are needed to obtain reliable data. As an essential part of the live cycle of data is the DataFlow crucial for acquiring high-quality data at the right time and location.
- **The need for SMART MetaData**: Columns of numbers of a time series alone are not useful without the context these numbers have been generated. The suitable description of data is a prerequisite for any secondary use of data. Apart from FAIR descriptions, the data trustworthiness also needs to be assessed and described to allow a correct evaluation of the data. Compiling this data in a complete manner and raising the awareness again, that MetaData are crucial for the correct use of data, is the real challenge for SMART MetaData.
- **The need for SMART Sampling**: Objectively finding the best possible sample location in space and time (most informative information for the respective research question), ideally in an automated and adapting way, is a challenging task. SMART sampling strategies are supporting this challenge. Applying state-of-the-art statistical and AI methods jointly with interactive visualization and analyses is increasing in the community. The challenge is to spread the knowledge about these methods and present easy ways of using them to lower the hurdle of their application.

Addressing these challenges was the main objective of the SMART Monitoring efforts within the Digital Earth project. The involved research centers started, iterated, and further developed the idea of an expanded SMART Monitoring Concept that finally integrates four conceptual groups of tools, each tackling one of the above-stated challenges.

2.2.3 Collaborative Interdisciplinary Working

Collaboration is essential for the success of the Digital Earth endeavor. Collaboration has to be managed on several levels: between various Earth science disciplines, between data science and Earth science, and between the involved research centers. We identified the following crucial issues in the project that we had to find solutions for:

- Establish topical working groups to shape a framework for collaboration across disciplines. For this, we defined two showcases: (a) the analysis of flood events at the Elbe River along the process cascade event generation, evolution and impact across atmosphere, and terrestrial and marine disciplines; and (b) quantification of methane emission fluxes into the atmosphere from gas exploration in the North Sea.
- Establish and implement digital collaboration platforms for information management and exchange. Mainly, we applied confluence for information sharing and GitLab for collaborative software development.
- Promote existing or upcoming infrastructures, agreements, and policies such as standards, licenses, or eScience infrastructures.

Chapter 7 presents a social science-oriented evaluation in which a World Cafe and a survey were used to evaluate the interdisciplinary collaboration and opportunities for improvement.

As Digital Earth is a pilot project, all process steps in collaboration, scientific workflow setup, method and tool development and hence scientific progress have been evaluated regularly throughout the project period using different measures (see Chapter 8). These evaluations during the project lifetime improved the process steps and produced an added value for the investigation of the Earth system and interdisciplinary collaboration.

To summarize, Digital Earth is designed as a pilot project that integrates data science methods, such as machine learning or visual data exploration into Earth and environmental science, and thus expands and enhances traditional analytical procedures. Digital Earth advances data science with the concrete application field in Earth science. Research in data science is necessary to tailor and enhance existing methods to the specific requirements resulting from Earth sciences. Furthermore, Digital Earth is a kind of a socio-cultural and organizational pilot project on collaboration between institutions and disciplines with a continuous evaluation of the progress.

In order to make the results of the described main topics of Digital Earth known to the communities of Earth sciences as well as data sciences, we have compiled them in this book. The aim of the book is to present the methods and solutions for overcoming the challenges of the three main topics in a compact way. In the following chapters, the book deals with the visual approaches (Chapter 3), the computational approaches (Chapter 4), and the developed scientific workflows (Chapter 5) of the data analysis and exploration. The collection of data using the Digital SMART monitoring approaches is described in Chapter 6. The concepts of interdisciplinary collaboration are conveyed in Chapter 7 and the evaluation of the Digital Earth approach for digitalization in Chapter 8. Finally, the lessons learned from the project are presented in Chapter 9.

Chapter 3
Data Analysis and Exploration with Visual Approaches

Everardo González Ávalos, Doris Dransch, Nicola Abraham, Valentin Buck, Daniel Eggert, Tom Kwasnitschka, Daniela Rabe, Flemming Stäbler, and Viktoria Wichert

Abstract A comprehensive study of the Earth system and its related processes requires a holistic examination and understanding of multidimensional data acquired with a large number of different sensors or produced by various models. To this end, the Digital Earth project developed a set of software solutions to study environmental data sets using visual approaches. In the following chapter, we present three data visualization products developed to deal with the challenges of the analysis and exploration of environmental data.

Keywords Data visualization · Data exploration · Spatiotemporal exploration · Linked views · Immersive visualization

3.1 Challenges

The incessant processes shaping our Earth's environment are determined by an interplay of diverse phenomena of physical, chemical, and biological nature, with ranges of action that span from planetary scales into the microscopic realm. As such, the study of the geoscientific processes and the interplay of determining phenomena rely on the analysis of highly diverse kinds of data from many sources. Several challenges arise from this situation:

1. The need to establish connections between causes and consequences of geoscientific processes. The mechanisms that link these processes together might be

E. González Ávalos (✉) · V. Buck · T. Kwasnitschka · F. Stäbler
GEOMAR Helmholtz Centre for Ocean Research Kiel, Kiel, Germany
e-mail: egonzalez@geomar.de

D. Dransch · D. Eggert · D. Rabe
Helmholtz Centre Potsdam - GFZ German Research Centre for Geosciences, Potsdam, Germany

N. Abraham · V. Wichert
Helmholtz-Zentrum Hereon, Geesthacht, Germany

© The Author(s) 2022
L. M. Bouwer et al. (eds.), *Integrating Data Science and Earth Science*,
SpringerBriefs in Earth System Sciences,
https://doi.org/10.1007/978-3-030-99546-1_3

few and scattered; thus, the connections become evident only when observations from various disciplines and sources are brought into relationship with each other.

2. The need to retain a sense of spatial and temporal coherence across different scales. The sense of spatial and temporal coherence is easily lost when regarding information across different scales simultaneously. This can be the case with sediment samples that encode information in cm scale, while remote sensing data does so in km^2 scale. In the temporal dimension, an underwater sediment plume can arise and settle in a matter of minutes, while global climatic phenomena are compared with each other across decades.

3. The need for suitable means to integrate a variety of heterogeneous spatiotemporal data sets. Scientists have to be supported in creating a "holistic view" on processes and related phenomena.

Digital Earth addresses these challenges with visualization. A main advantage of visualization is its ability to simultaneously display data even when it differs in scale, variables, or accuracy. We use interactive visualization to display heterogeneous data sets in a unified 4-dimensional environment and to interactively explore the data with respect to context and connections.

We applied different visualization techniques and environments and adapted them to our geoscientific requirements: We developed and respectively utilized following visualization tools:

– The Data Analytics Software Framework (DASF) which provides linkable visualization components (multiple linked views),
– The Digital Earth Viewer, an engine for 4D data contextualization and visualization, and
– The ARENA2, an immersive visualization infrastructure.

3.2   The Data Analytics Software Framework (DASF) Providing Linkable Visualization Components

3.2.1   Introduction

The Data Analytics Software Framework (DASF) (Chapter 5.2.3) which we have developed in Digital Earth aims at implementing scientific data analysis workflows. Besides the module to integrate components into a scientific data analysis workflow, it also provides a visualization module to present the data and results that are used and created in the workflow.

3.2.2 Visualization Concept

In order to enable scientists to simultaneously show and explore the data in its multiple dimensions (space, time, different variables, accuracy), the DASF visualization module consists of a variety of visualization components that are linked according to the multiple linked view visualization approach (Roberts 2005; Spence 2007). The visualization components can be all types of views on data: maps, diagrams, tables, animations, or calendar maps to name a few. The visualization components are presented in several windows simultaneously. The windows are linked; this means that operations in one window affect all other related windows. For example, if a user selects a subset of data in one window (e.g., in a map), all other windows visualize information for the same selected data subset. The operations are executed interactively with mouse-based interaction techniques, such as brushing, highlighting, or filtering. Multiple linked views are a widely used technique in data and information visualization to present multivariate and multidimensional data sets.

3.2.3 Technical Implementation

Our architecture to technically implement the DASF visualization module utilizes well-established techniques. To create the single visualization components, we used for instance *Openlayers,* (https://openlayers.org) or *leaflet* (https://leafletjs.com/) for maps, *D3* (https://d3js.org/) or *chartjs* (https://www.chartjs.org/) for charts (e.g., histogram, time diagram, radarplot), and *Vueftify* (https://vuetifyjs.com/en/compon ents/data-tables/) for tables. To implement the multiple linked view approach, we applied the "reactive properties" concept which is provided by the *Vuejs and Vuetifyjs* software package (https://vuejs.org/v2/guide/reactivity.html). For more information on the complete DASF implementation see, Chapter 5.2.3.

3.2.4 Application

The DASF visualization module has been applied in all workflows contributing to the Digital Earth Flood Event Explorer (Chapter 5.3). For each workflow, an aligned visual interface was assembled from the various visualization components and the multiple linked view approach. Exemplarily, we show the visual interface of the "River Plume Workflow" which supports geoscientists to investigate the impact of river floods on the marine environment (Chapter 5.3.3). It should enable scientists to detect the spatiotemporal influence of the river plume on the sea due to chemical anomalies and to answer the question: Where and when can we detect the flood river plume in the sea? Several data sets have to be used and combined to answer this question. These include observations of chemical characteristics of the waterbody, such as salinity, which are collected by a sensor at regular intervals along a ferry route, and a data set from a physical model that calculates model trajectories of the waterbodies observed on the ferry route up to 10 days into the past and 10 days into future from a reference day. On the basis of these data sets, anomalies of salinity, chlorophyll or surface temperature, and thus the spatiotemporal behavior of the river plume can be detected. Chlorophyll anomalies related to the river plume occur when the deviation is above the expected range, while salinity and surface temperature anomalies occur when the deviation is below the expected range.

The multiple-view visualization consists of following components (Fig. 3.1): The map view (V1) shows the spatial distribution of the observed ferry box data and calculated model data; the color encodes the concentration of a chemical or physical parameter such as salinity. Another view presents a comparison of the quantities for each chemical or physical parameter measured inside and outside of a user-defined region of interest. The bar chart (V3) gives an overview of all chemical/physical parameters; the table below (V4) shows detailed information for one selected parameter. A further view (V5) presents temporal information and visualizes the occurrence of anomalies of the parameters in time in a calendar-heatmap. These anomalies are

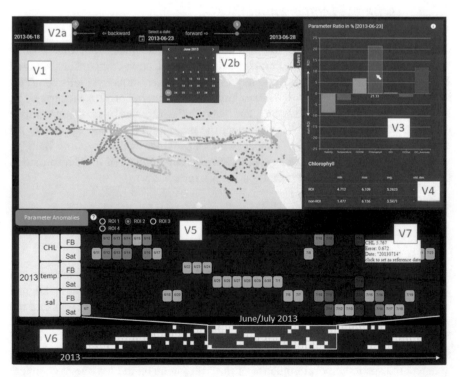

Fig. 3.1 Multiple linked views to determine the spatiotemporal behavior of the flood's river plume in the waterbody, such as the North Sea (Interface-in-Action Video: https://youtu.be/yl8ngubBxYY)

candidates for detections of the river plume in the observational data and are determined automatically through a Gaussian Regression algorithm (Chapter 4.5.1). In
the calendar-heatmap, days with deviations from the expected range of each variable are shown with different color intensity. More details about the anomaly can be
added for each entity in the heatmap through a mouse-over action to load the relevant
observational and model data into the interactive map (V7). An additional overview
presentation (V6) shows the whole data set and puts the data subset presented in V5
into context. The links between the single views are realized with interactive operations like the filtering of a region of interest (V1), the selection of a continuous time
interval (V2a), or discrete time step (V2b), or by mouse-over actions for presenting
additional information (V7).

Added Value: The visual interface of the "River Plume Workflow" supports scientists to visually put the various data sets into context: anomalies of chemical/physical
parameters and their spatial and temporal distribution. The overall view on the one
hand and the capability of interactive data exploration on the other hand assist scientists to finally detect the river plume and its behavior in space and time. The combination of the visual interface and the Gaussian Regression algorithm to detect the
river plume that was also developed in Digital Earth (Chapter 4.5.1) provides a novel
approach for scientists to comprehensively analyze the various data sets and to detect
the river plume.

3.3   The Digital Earth Viewer

3.3.1   Introduction

The Digital Earth Viewer is a web application for spatiotemporal contextualization
and visualization of heterogeneous data sources. It was developed with the goal of
enabling real-time exploration of geoscientific data sets across spatial and temporal
scales. To this end, it is capable of ingesting data from a large variety of types that
are usually found in the geosciences, and it deploys a user interface, which allows
for interactive visual analysis. At the same time, online and offline deployment,
cross-platform implementation, and a comprehensive graphical user interface are all
capabilities that make the Digital Earth Viewer particularly accessible to scientific
users.

3.3.2   Visualization Concept

This infrastructure provides a framework in which new visualizations for heterogeneous data set can be created with relative ease. Since no reduction in dimensionality

is undertaken and the four-dimensional data (space and time) is displayed in a four-dimensional context, temporal or spatial distortions are mostly avoided; this leads to an improved interpretation capability and supports the understanding and contextualizing of information in an intuitive process. The Digital Earth Viewer enables the user to visualize assorted geoscientific data in three spatial dimensions and across time. It projects the spatial coordinates latitude, longitude, and altitude into a virtual globe and builds an ordered registry of temporal events. Both of these features can be accessed through user interface elements in real time.

Different data sources can be added simultaneously to form individual layers, each with its own data basis and transformation pipeline. Global parameters, such as the position in time and space or the intrinsic scale of the visualization, are implicitly and explicitly communicated to the user in the graphical interface while specialized parameters can be set through a menu. The transformations for each data source happen independent from one another and are composed together into one final result, allowing the blending of multiple data sources.

Data is grouped into several different categories for display. Traditional 2D maps can be projected onto a spherical surface and displacement along the sphere's normal vector can be applied. Scalar values are mapped onto one of a set of color maps, while precolored maps are passed through. Sparse data can be displayed as a point cloud which is projected, colored, and culled according to the global projection parameters. For the intuitive representation of vector fields, an animated particle system is created in which the particles follow the vector field which is projected onto the virtual globe.

3.3.2.1 Technical Implementation

The tool is a hybrid application, which is split into a server back-end and a client front-end. The rust (https://www.rust-lang.org/) back-end handles data extraction from different file formats as well as re-gridding into viewable areas and caching. It can be hosted remotely on a server or locally for offline access. The front-end consists of an HTML interface component and is responsible for the 3D data rendering using the WebGL API (https://www.khronos.org/webgl/) and the implementation of the graphic user interface controls which uses Vue.js (https://vuejs.org/).

For each data type that the server needs to ingest, a specific data type is built that transforms the incoming data format into an internal representation optimized for computation operations. This representation is then passed over to the client, which applies graphical transformations to compute a visual representation of it.

A running instance of the application can be accessed under the following web address: https://digitalearthviewer.geomar.de.

The Digital Earth Viewer is an open-source software licensed under the EUPL (https://joinup.ec.europa.eu/collection/eupl/eupl-text-eupl-12).

3.3.3   Applications

3.3.3.1   Methane Budget in the North Sea

The Digital Earth showcase "Methane Budget in the North Sea" set up to build a methane gas budgeting for the North Sea region. The showcase makes use of the Digital Earth Viewer to unify a large number of data sets under a single visualization interface. Boreholes from fossil fuel production are known to be important sources of methane that is released into the atmosphere. The GEBCO (https://www.gebco.net) bathymetry from the North Sea region is displayed in a 3D elevation model. The aerosol and trace gases dispersion model ICON-ART (Rieger et al. 2015) is used to calculate the contribution of these boreholes to the atmospheric methane concentration. The viewer's interface allows to quickly compare the resulting data product with existing methane estimates from the EDGAR (https://data.jrc.ec.europa.eu/col lection/edgar) emissions database and provides a visual assessment of their accuracy. In a similar way, measurements of geochemical water properties from the expedition POS 526 (https://oceanrep.geomar.de/47036/) are displayed in spatial context of other measurement compilations from the Pangea (https://www.pangaea.de) and MEMENTO (Bange and Bell 2009) databases. The observation of their development over time is further supported by the visualization of three-dimensional physical water properties like current velocities and pycnocline depth obtained from the NEMO Model (Madec 2016). An instance of the Digital Earth Viewer displaying the Methane showcase can be found under following web address: https://digitalea rthviewer-methane.geomar.de.

Added value: Using the Digital Earth Viewer, scientists can simultaneously access and visualize data from all the sources mentioned above. Seamless spatial navigation allows them to directly compare the global impact that regional methane sources have in the atmosphere, while temporal components enable them to do so across the different seasons of an entire year (Fig. 3.2).

3.3.3.2   Explorable 4D Visualization of Marine Data

The expedition Mining Impact-II started a series of experiments to answer some of the most important questions regarding profitability and sustainable exploitation of resources in deep sea mining, such as mining for manganese nodules. Deep sea exploration is a challenging endeavor that can be greatly aided by the use of modern visualization techniques. In this work, we aim to recreate a sediment plume which resulted from an underwater dredging experiment. This will help to quantify similar sediment depositions from mining that could impact deep sea ecosystems at a depth of 4,000 m. Sensor data fusion allows for the virtual exploration of experiments on the seafloor; the following is an overview of the different data sources acquired during the expedition that come together within one visualization:

Fig. 3.2 Digital Earth Viewer used to display the North Sea area, atmospheric methane calculations from the ICON-ART model, and methane flows from oil and gas wells

- Turbidity sensors calculate this optical property of water by measuring the scattered light that results from illuminating the water.
- Current sensors use the Doppler effect to measure the velocity of particles suspended in the water column and thus calculate the speed and direction of water currents.
- Multi-beam echosounders emit fan-shaped sound waves and use time of flight to reconstruct the seafloor bathymetry.

For the dredging experiment, an array of 8 turbidity sensors and 8 current sensors was placed on the sea floor in an area previously scanned with a high-resolution multi-beam echosounder mounted beneath an autonomous underwater vehicle. Moreover, a virtual sediment plume was modeled and integrated into the experiment. The spatiotemporal contextualization of all data sources took place allowing for a real-time simultaneous analysis of heterogeneous data sources in 3D and across time. An instance of the Digital Earth Viewer displaying the Sediment Plume showcase can be found under following web address: https://digitalearthviewer-plume.geomar.de.

Added value: Using the Digital Earth Viewer, the experiment grounds were recreated. This virtual environment allowed scientists to peer beyond the darkness of the deep sea and explore the impact of a simulated mining endeavor. The numerical model of the sediment plume was superimposed and compared to the in situ data obtained by the turbidity and current sensors resulting in a visual confirmation of the sensor placement and model correctness. Deployment as a web-based application accounted for cross-platform portability across devices and operating systems, allowing scientists to visually explore the phenomena that take place in this virtual abyssal plain, and to share their discoveries with a wider audience (Fig. 3.3).

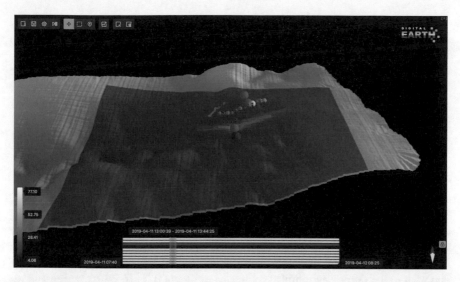

Fig. 3.3 Parallel exploration of multiple marine data types: the Digital Earth Viewer is used for the visual corroboration of a 3D plume model. This is done comparing the plume dispersion values (small orange dots) with the readings of the turbidity sensors (larger green and red dots) and the water currents (green lines)

3.4 Spatially Immersive Visualization of Complex Seafloor Terrain

3.4.1 Introduction

To a large extent, field geologists derive their mental models of complex outcrops and depositional features (e.g., volcanoes) from situational awareness and the first-person perception of an environment via the bodily senses. To date, this is still reflected in the way geologists are trained, even in light of emerging digital technologies. Moreover, economies of scale (e.g., geospatial correlation of small outcrops) unfold only across large "industrial scale" survey areas, as opposed to isolated, confined demonstrator missions. In order to bridge the scales of these two different planes of operation and to retain a true sense of dimensionality while doing so, the ARENA2 spatially immersive visualization laboratory was developed at GEOMAR.

3.4.1.1 Visualization Concept

The ARENA2 projects geospatial data onto an elevated dome and allows one or multiple users to navigate freely across all three dimensions of a virtual environment. A faithful recreation of live exploration of the geological features is enabled by this.

Notably, most of the ARENA2 applications are also available in desktop environments, or even through web browsers. This creates a continuum of visualization infrastructure scaling from opportunistic, personal access all the way to collaborative, structured visualization campaigns serving the entire spectrum of academic use cases: data sets that were previously cured and preprocessed using commonly available software applications on a desktop PC can be displayed and analyzed by multiple users with a new sense of immersion.

3.4.2   Technical Implementation

The ARENA2 features a tilted, stereoscopic projection dome covered by a five-channel projection system, which is in turn fed by a node-based visualization cluster of five computers. We follow a three-tiered approach on visualization software: First, the graphical output of the photogrammetric post processing software itself is ported to the dome environment by means of OpenGL buffer distribution and re-interpretation across the cluster, known as openGL-hooking. Second, a dedicated, distributed visualization software loads a statically exported point cloud in a georeferenced virtual globe context. Third, parallel WebGL-based visualization tools are synchronized across the cluster, with dome-specific warping and blending applied on the operating system level. Here, a preprocessed level-of-detail pyramid of the point cloud is dynamically streamed to all web clients.

3.4.3   Application

To exemplify the procedure, an inaugural data set demonstrator was created: We surveyed a 500×500×80m volcanic crater in 2016, hosting an extensive, active hydrothermal field at the Niua South Volcano, Tonga, by means of the remotely operated vehicle (ROV) ROPOS (https://www.ropos.com/index.php/ropos-rov). During 100 hours of a methodical survey, some 220.000 photographs of the seafloor were collected using a single-lens reflex camera and transformed into a 3D, color textured terrain model. Its scale and detail form a considerable challenge for agile, interactive visualization procedures.

Added value: Scientists studying these hydrothermal vents will be able to do so in life-like detail. The virtual environment can be freely explored without the physical constraints of the ROV used for the original acquisition. For the wide majority of the population that will never be able to take part in an expedition such as the one where these geological features were captured, the ARENA2 can help bring these people closer to the richness of the underwater landscapes with an immersive experience (Fig. 3.4).

Foto: Nikolas Linke / GEOMAR

Fig. 3.4 ARENA2 projection dome displaying a bathymetric relief of the deep sea. The infrastructure provides an immersive experience for the presenter and the audience alike, opening new narrative possibilities and discussion planes

3.5   Assessment of the Three Visualization Approaches and Techniques

The three data visualization techniques developed in the Digital Earth Project respond to the challenges formulated in 3.1. All integrate a variety of heterogeneous spatiotemporal data sets, but each does so in a different way. The linked views of the DASF relate directly to the first challenge and propose an efficient answer with the use of graphs and charts for data conceptualization. These provide rapid access to insights of interlaced behaviors which a user can then recognize as causal links. The 3D data representation capabilities are the core strength of the Digital Earth Viewer which excels at exploring all dimensions of spatiotemporal data from multiple sources. It addresses specially the second challenge, which is to retain a sense of spatial and temporal coherence across different scales. Both the applications based on DASF and the Digital Earth Viewer are deployed as web applications and thus are accessible across platforms. Considering the ARENA2 consists of a building-sized infrastructure, accessibility and deployment are by far its largest limitations. To make up for this, it provides unparalleled data immersion capabilities and a revolutionary way of experiencing data, during both the exploration and presentation of geoscientific data sets. With this, it also addresses challenges 1 and 2.

References

Bange HW, Bell TG (2009) MEMENTO: A marine methane and nitrous oxide database, Solas News, Issue 9, Spring 2009, p 42

Madec G, NEMO System Team, 2016: NEMO ocean engine. Scientific Notes of Climate Modelling Center (27) - ISSN 1288–1619

Rieger D et al (2015) ICON–ART 1.0—A new online-coupled model system from the global to regional scale. Geosci Model Dev 8:1659–1676

Roberts JC (2005) Exploratory visualization with multiple linked views. In: Dykes J, MacEachren AM, Kraak M-J (eds) Exploring geovisualization. Elsevier, pp 159–180

Spence R (2007) Information visualization. Design for interaction. Pearson

Chapter 4
Data Analysis and Exploration with Computational Approaches

Viktoria Wichert, Laurens M. Bouwer, Nicola Abraham, Holger Brix,
Ulrich Callies, Everardo González Ávalos, Lennart Christopher Marien,
Volker Matthias, Patrick Michaelis, Daniela Rabe, Diana Rechid,
Roland Ruhnke, Christian Scharun, Mahyar Valizadeh, Andrey Vlasenko,
and Wolfgang zu Castell

Abstract Artificial intelligence and machine learning (ML) methods are increasingly applied in Earth system research, for improving data analysis, and model performance, and eventually system understanding. In the Digital Earth project, several ML approaches have been tested and applied, and are discussed in this chapter. These include data analysis using supervised learning and classification for detection of river levees and underwater ammunition; process estimation of methane emissions and for environmental health; point-to-space extrapolation of varying observed quantities; anomaly and event detection in spatial and temporal geoscientific datasets. We present the approaches and results, and finally, we provide some conclusions on the broad applications of these computational data exploration methods and approaches.

Keywords Machine learning · Artificial intelligence · Earth system · Data exploration

V. Wichert (✉) · N. Abraham · H. Brix · U. Callies · V. Matthias · A. Vlasenko
Helmholtz-Zentrum Hereon, Geesthacht, Germany
e-mail: viktoria.wichert@hereon.de

L. M. Bouwer · L. C. Marien · D. Rechid
Climate Service Center Germany (GERICS), Helmholtz-Zentrum Hereon, Hamburg, Germany

E. González Ávalos · P. Michaelis
GEOMAR Helmholtz Centre for Ocean Research Kiel, Kiel, Germany

D. Rabe · W. zu Castell
Helmholtz Centre Potsdam—GFZ German Research Centre for Geosciences, Potsdam, Germany

R. Ruhnke · C. Scharun
Karlsruhe Institute of Technology, Eggenstein-Leopoldshafen, Germany

M. Valizadeh · W. zu Castell
Helmholtz Zentrum München—German Research Center for Environmental Health, Neuherberg, Germany

L. M. Bouwer et al. (eds.), *Integrating Data Science and Earth Science*,
SpringerBriefs in Earth System Sciences,
https://doi.org/10.1007/978-3-030-99546-1_4

4.1   Introduction and Challenge

Computational data exploration and analysis can help to substantially improve modelling and understanding of Earth system processes. In this chapter, we provide an overview of the developments in the Digital Earth project that focus on employing such innovative techniques to improve our process understanding, to derive new insights from a variety of existing datasets and to make the investigation of complex processes more feasible. Diverse sub-disciplines in the Earth sciences are using computational methods to solve some of the major issues identified for the Earth science community. The issues for which computational applications have been developed in Digital Earth and that are presented in this chapter are as follows:

- Extracting relevant information and features using machine learning approaches: For various features, no labelled data collections exist, as these are too labour intensive to develop. Labels are important for supervised learning algorithms, for example, to classify specific observations using prior knowledge. Using sparse datasets and machine learning methods, alternative ways can be found to broaden data availability and derive new, crucial information from existing data. Here, examples are provided that map river levees in Germany, for which no consistent data were readily available for research before and for detecting underwater ammunition locations.
- Approximating complex processes with machine learning: Although models are successfully used to understand complex processes in the Earth system, for some applications the computational cost is too high to embed this information in a broader framework and to answer questions that are more challenging. In these cases, approximating these same processes through machine learning algorithms can be a means for scientists to tackle those problems. We present an approach to estimate methane and ethane concentrations through a Neural Network and give an example of how machine learning algorithms can be used to combine highly heterogeneous data to answer pressing questions related to climate change and health research.
- Point-to-space extrapolation: For many applications in the Earth sciences, single-point observations need to be inter- and extrapolated across space to arrive at consistent estimates of total matter fluxes. This chapter presents an example where point observations of methane emissions are analysed and processed in order to be consistent with global atmospheric emissions as observed/estimated in global databases. A second application provides insights into the functionalities of advanced approaches for point-to-space extrapolation.
- Anomaly and event detection across heterogeneous datasets: In some applications, process understanding can be improved considerably when data from diverse sources are combined through computational methods. Detection of events and anomalies is important in Earth systems for scientific and for practical applications. We present an approach that combines observational and model data to detect river plumes at sea at the end of a riverine flood event chain and tracks their spatial and temporal extent.

4.2   Object Recognition Using Machine Learning

4.2.1   Deep Learning Support for Identifying Uncharted Levees in Germany

Germany has a large network of levees to manage flood events. Unfortunately, data on the locations of these levees are not always directly available to the public or to researchers. Due to the importance of these levees in the analysis of flood events, approaches are needed to derive the levees' location and height from available information. However, such methods are not readily available, and neither are commonly accepted nor standardized approaches. Advances in computational methods, namely deep learning, and the release of a wide range of geodata to the public make it possible to find levees automatically and on a large scale.

In this research, we have started to develop such a framework and appropriate methods to delineate levee features from such data. As data sources, we combined aerial images and LIDAR-based digital elevation models. The raster format of this data is comparable to image data, and as such, we apply deep learning methods, which is providing state-of-the-art solutions for various computer vision problems. For example, in medical image analysis, deep learning models are used for cell classification and the tumour detection.

Our approach relies on semantic segmentation, a common task in deep learning. Semantic segmentation refers to the classification of individual pixels to different predefined classes. The output has the same raster shape as the input features. To train a semantic segmentation model, a mask with a classification of the input features is needed. In our case, these input features are the pixels from the aerial images and digital elevation model. We choose a common architecture for our semantic segmentation model, the U-Net (Ronneberger et al. 2015). A U-Net consists of blocks of convolution layers in combination with pooling (to reduce the data size by a factor of two) or upscaling (to increase the data size by a factor of two) layers. There is an equal number of pooling and upscaling blocks. The pooling blocks come first, followed by the upscaling blocks, transferring the information from input to output. Additionally, the output of the first pooling blocks is used as an input for the last upscaling block, and the output from the second pooling block is used as an input for the second to last upscaling block. This schema continues for the other blocks in the network.

To train our model, we used the publicly available data from North Rhine-Westphalia (NRW 2021), which contains the relevant information. As features, we use the LIDAR-based digital elevation model (1 m resolution in two by two-kilometre tiles) and the aerial photographs (0.1 m resolution in one by one-kilometre tiles). The levees are available as shapefiles for the crest of the levees. Several pre-processing steps are applied.

The first step is to rescale the aerial images to the resolution of the digital elevation model and to split the digital elevation model into one by one-kilometre tiles. Afterwards, we rescale the pixel values for all inputs to a range of zero to one. The processing of the labels is more complex, as we need to create a mask from the line shapes. The lines themselves only cover a small subset of the pixels; therefore, we use the entire width of the levees, which is not given in the dataset. To derive the width of a levee, we take orthogonal sections to each line segment and look for the maxima in the second derivative of the digital elevation model along the sections. We use these maxima as the boundaries of the levees. In addition to the derived masks for levees, we also include a mask for the bodies of water which is derived from publicly available polygon shapes. The training itself is run on GPUs and includes common data augmentation techniques such as rotations.

The results must be in the same format as the original input, so we have to process the output of the neural network and extract the information. This post-processing consists of multiple steps. The first step is to assign contiguous areas of pixels to different groups. This is based on a threshold value as the neural network output is the probability of a pixel to be of one class. We also apply a maximum filter of size four to remove some noise which can be induced by applying the threshold. For each group of pixels, we then look for the highest points as we want to find the crest of the levee. These points should all fall in the same height range, a criterion we use to exclude unlikely levees. The next step is to merge groups of adjacent tiles to get the entire levee as one object. Additionally, we analyse the length of the crests and discard short sections. All points of the merged groups are then transferred to a line shape. The shape is simplified using standard tools to reduce the number of points specifying the shape. These output shapes can then be used for analyses. We additionally use the pre-processing method to create polygons for the entire levees (c.f. Fig. 4.1). Overall, the methodology detects a high percentage of the charted levees, where

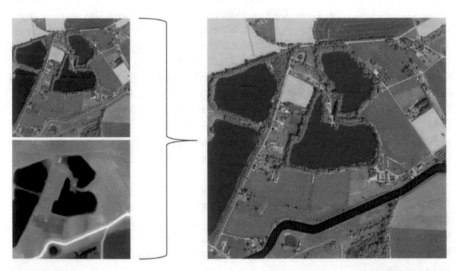

Fig. 4.1 Exemplary output of the deep learning model after post-processing. The input data (aerial image on *top left* and digital elevation model at the *bottom left*) together with the output, the aerial image with the predicted shapes as overlays

precision and recall can be balanced by adjusting the selection threshold. The next step is to evaluate the methodology and model using data from a different state, e.g. Saxony. In general, this approach is applicable to a wide range of problems in remote sensing. The fusion of different data sources is still uncommon in combination with deep learning models. Our use case highlights the benefit of such an approach.

4.2.2   Machine Learning Support for Automated Munition Detection in the Seabed

Both the North Sea and Baltic Sea have been used as dumping grounds for munitions, especially after the Second World War. Nowadays, various infrastructure projects are planned and built in these waters, including offshore wind farms and pipelines. Before construction can begin, the area must be cleared of munitions. Multi-beam echo sounders, side-scan echo sounders, magnetometers and sub-bottom profilers are used to explore the construction sites. It is, however, very time consuming to analyse the data generated by the instruments. We propose the use of machine learning to detect munitions in the data.

For our analysis, we rely on the multi-beam echo sounder. The echo sounder dataset provides the water depth and the backscatter, giving information on the composition of the seafloor, at a horizontal resolution of 25 cm (where depths are below 20 m). The integration of other data sources, e.g. magnetometer data, into one model is possible, but comes with a number of challenges. The most important one is getting an accurate spatial alignment of the datasets. A mismatch of even one metre can cause many objects not to overlap in the combined dataset.

The machine learning method we applied is deep learning. As with the previous application for levee detection, we face a semantic segmentation problem. The model class we use to approach this problem is a U-Net (Ronneberger et al. 2015). After testing various model configurations with depth (number of layers) and width (number of filters in a layer), we can conclude that the problem can be solved with relatively small models (with respect to both the depth and width). These models can be trained on the CPU within less than an hour.

To create a training dataset for our machine learning model, a few steps are necessary. First, we need a labelled dataset to be able to train a model. The second step is the preparation of the labels for the model type. In our case for multi-beam echosounder data, we need to create a mask for the map, based on available point labels. We achieved this by labelling an area around the point label as targets. This way we sometimes falsely label data as targets, but also account for imprecisely placed labels. To facilitate the training of a model, we scale and standardize the data in terms of the range of values. Here, we keep track of the exact steps and parameters used to be able to apply the model and processing steps to other datasets in the future. Additionally, we use standard machine learning practices like train-test splits and augmentation techniques like rotations and mirroring.

Fig. 4.2 Prediction of the neural network on top of the backscatter. *Red areas* indicate a high predicted probability for munitions, and clear areas indicate a low predicted probability. The *green dots* are the original labels

The results of the model (Fig. 4.2) look very promising as we can detect most of the labelled objects (>95%). However, we get several false positives, at least according to the labels, which might not be complete. The second important consideration is that the labels are not validated and might be false. Therefore, the number of false positives might be even higher.

We transferred this approach to a sub-bottom profiler. In both cases, we have two-dimensional data. The difference is that the data in this case are not in latitude and longitude direction but along a transect in latitude and longitude with the depth as the second dimension. One observation is that the models must be more complex to get good results. An important difference is in the training process where rotations are not a viable augmentation technique because objects create a distinct bell-shaped curve in the sub-bottom profiler images. The training can be done on a CPU as well. Overall, the issue of false positives persists while most labelled objects are found.

4.3   Approximating Complex Processes with Machine Learning

4.3.1   Estimation of Methane and Ethane Concentrations in the Atmosphere Over Europe by Means of a Neural Network

Methane is an important atmospheric greenhouse gas that has a substantial impact on climate and air quality (Van Dingenen et al. 2018). Chemical transport models (CTMs) are the most widely used tools allowing us to predict its concentration in air and its possible effects on the environment. A typical CTM accepts emissions and meteorological data as input data and calculates the concentration changes of atmospheric methane (and other gases) in time. Although CTMs have been continuously improved, they require a significant amount of computational resources (CPU time, RAM and disc space). Neural networks may become a cheaper alternative to CTMs in terms of these resources. The idea of a neural network (NN) is to fit a combination of simple mathematical functions (neurons or activation functions) so that for a given set of predictors and predictands, a NN, receiving predictors as inputs, estimates outputs that have minimal difference with the corresponding observed predictands. After training, the NN should be able to predict unknown output based on any given input. To verify its predictive skill, the NN is tested on an independent set of known inputs and outputs that were not employed in training.

To estimate methane concentrations, we developed two neural networks. Note that 19% of atmospheric methane is associated with fossil fuel production, mainly related to oil and gas mining (Van Dingenen et al. 2018); more than half of it leaks directly from the gas or oil fields. To detect these leakages from the offshore fields, we developed the first NN that estimates local anomalies in methane concentration directly from measurements near the potential (or known) natural gas source. We built this NN using Keras/Tensorflow package (Abadi et al. 2016). It consists of three dense layers with a hyperbolic tangent activation function, eight inputs (latitude, longitude, the temperature at two-metre height, time, humidity, latitudinal and longitudinal wind components at ten metres, sea surface temperature) and one output (methane). The NN was trained on the cruise measurements POS-526 (Greinert and Schoening 2019) that took place from 07.23.2018 to 11.08.2018 on a route: Bergen (Norway)—Dogger Bank (Netherlands)—Hirtshals (Denmark)—Tisler (Norway). The cruise data contain all input and output variables. Note that having geographical coordinates as inputs, the NN "learned" during the training the positions (and the corresponding emissions) of wells and oil fields by fitting its estimates to the methane concentration at different locations in the training data. After training, this NN, installed on a laptop, can estimate methane concentration at the current location from the current physical parameters of the surrounding atmosphere. If this estimate does not match the measured concentration, one may suspect to have detected an anomaly possibly associated with new oil or gas fields, or substantial changes in the known ones.

Note that measured methane anomaly may originate from other sources. To exclude the impact from other sources, we developed a second NN, which estimates daily mean ethane concentration anomalies in the atmosphere. Natural gas contains up to several per cent of ethane, giving 62% of atmospheric ethane (Franco et al. 2016). The concentration ratio of ethane/methane is unique for each oil or gas field and constant in time, serving as a kind of fingerprint (Visschedijk et al. 2018). Since the methane/ethane ratios near oil or gas fields are known (Yacovitch et al. 2020), we can estimate the fraction of atmospheric methane leaking from the gas or oil fields just from ethane. The second NN was developed on the basis of the network described in (Vlasenko et al. 2021). We trained and tested the second NN on the ethane anomalies estimated from the Consortium Multiscale Air Quality Model (CMAQ) (Appel et al. 2013) in the European domain for the period 1979–2009 using the same emission data for the year 2012 (Bieser et al. 2011) for the entire 30-year period. We define ethane anomalies as the deviation of the current value from its climatological mean. For the second NN, we again used the Keras/Tensorflow package (Abadi et al. 2016), choosing a NN that consists of one recurrent layer followed by two dense layers. It accepts wind anomalies in the European domain and estimates the corresponding ethane anomaly in the same area. All layers have hyperbolic tangent activation functions.

To train and test the NN, we split the data for both NNs as follows. The training set for the first NN contains 90% of samples, which is 14,400 inputs and outputs, randomly picked up from the POS-526 cruise measurements (Greinert and Schoening 2019). The remaining 10% of data, i.e. 1600 samples, were used for testing. For the second NN, we took estimated wind and ethane anomalies from 1979 to 2006 for training and anomalies from 2007 to 2009 for testing. Developing the second NN, we found that the inter-seasonal variability in the data resulted in errors. To minimize these errors, we created and trained the NN for each season separately.

To evaluate the accuracy of both NNs, we used the R^2 measure that shows how much of the observed data variability is explained by the model (which is the NN in our case). Note that for the second NN, anomalies obtained from CMAQ estimates play the role of observations.

The estimates of methane and ethane obtained from the first and the second NN are shown in Figs. 4.3 and 4.4, respectively. Note that the estimates of the first network (blue line) and the measured methane concentration (red line) almost coincide. As a result, R^2 for the first network equals 0.91. The R^2 for the second NN equals 0.565, 0.48 and 0.57 for summer, spring and winter, respectively. Although the second NN has lower R^2 than the first NN, it reconstructs the ethane anomaly patterns' main features. This can be seen in comparison with the summer mean ethane anomalies estimated by the NN (Fig. 4.4, left panel) and CMAQ (Fig. 4.4, right panel). Note that except for small details, the NN succeeds in reconstructing the pattern of the CMAQ simulations. Other deteriorations of R^2 are caused mainly by a slight underestimation of the anomaly amplitude. Relying on these results, we conclude that the first and the second neural networks predict the corresponding methane and ethane concentration anomalies with a high degree of accuracy and can be used as a smart monitoring tool during the researcher campaigns. Combining these neural networks into one

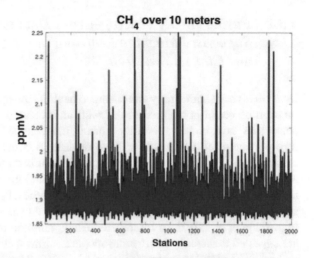

Fig. 4.3 Estimated (*blue*) and observed (*red*) methane concentrations, corresponding to the measurements in cruise POS-526 in the North Sea

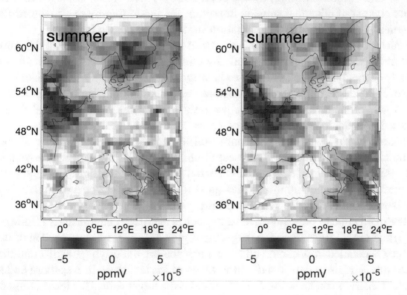

Fig. 4.4 Mean summer ethane concentration anomalies estimated with CMAQ (*left*) and NN (*right*)

predictive system, enabling the more accurate determination of the source of methane leaks is the next step in their development.

4.3.2 Fusing Highly Heterogeneous Data to Facilitate Supervised Machine Learning in the Context of Health and Climate Research

Heat waves can significantly affect human health. Examples include increased transmission of vector-borne diseases as well as increased susceptibility to metabolic conditions and higher mortality during episodes of severe heat. It is therefore paramount to investigate climate change in terms of potential-related health outcomes. To that end, the focus of our research is on temperature extremes, such as encountered during heat waves, and myocardial infarction (MI). Data from the region of Augsburg, Germany, are used as a case study. Epidemiological studies have shown that temperature extremes may indeed lead to an increased occurrence of MI (e.g. Chen et al. 2019). In future, frequency, duration and intensity of heat waves are expected to increase due to anthropogenic climate change, even at levels limited to $1.5°$ or $2°$ global warming (Sieck et al. 2020). Therefore, assessing health risks in the context of climate change is important for supporting more climate-resilient societies, public planning and adaptation strategies for human health.

Machine learning (ML) is a powerful tool for investigating complex and unknown relationships between environmental conditions and their adverse impacts and has already been applied in other fields (e.g. Wagenaar et al. 2017). ML is a data-driven approach, and meaningful results depend on consistent and high-quality data. To investigate climate change and MI, not only climate and meteorological data are required, but confounding effects of other well-known risk factors must also be accounted for by additional environmental, demographic, behavioural and socio-economic data. This makes this research challenging, as availability, provenance, and detail or resolution (temporal and spatial) of the data are highly variable. Here, we present a dedicated approach, designed to fuse such heterogeneous data into a consistent input dataset for ML algorithms.

The main pillar of our data-driven approach is the KORA cohort study (Holle et al. 2005) and the MI Registry in the Augsburg region of Bavaria, Germany. This dataset comprises detailed information on MI occurrence and underlying health conditions. Based on the registry data, a daily time series of MI incidence in Augsburg and two adjacent districts can be derived. This provides the target values for the training data for the ML algorithms.

To learn about the association of MI and the diverse risk factors such as exposure to heat, demographic structure, confounding health factors and air quality, these must be provided as predictors for the algorithms. Currently, weather and climate data (temperature observations and climate projections from the EURO-CORDEX initiative see Jacob et al. 2014); air pollution data (e.g. PM_{10}, PM2.5, nitrogen oxides and ozone; from regional environmental governments, such as BLFU, 2021); distribution of green spaces based on NDVI; demographic characteristics (age and sex, from regional statistical agencies, such as BLFS 2021); pre-existing illnesses (e.g. diabetes and obesity as recorded in the MI registry data); and socio-economic data (e.g. household income, education) are planned to be used within the project.

The raw data for these predictors are extremely heterogeneous for many reasons. First, the data come from different providers and is presented in various file formats, some of which are proprietary. Second, the representation of data can differ as well. Some data are of gridded/raster type (e.g. NDVI), some are point data (weather observations, air quality) and some are time series data, aggregated at the district level (e.g. demographic data). Third, the spatiotemporal scales (e.g. regional vs. local, coverage in time) and resolutions differ substantially.

The MI registry data are given in the form of individual cases with information on the date of the MI, the district where it happened and additional information on patient health. In a first step, a daily time series of MI incidence, aggregated at district level, is derived. The procedure is designed to produce compatible time series from the predictor data while addressing the three dimensions of heterogeneity outlined above. Afterwards, ML algorithms can readily be used to learn the relationship between incidence and the various predictors.

Figure 4.5 shows a schematic of the fusion procedure. Different predictors enter the pipeline and are subjected to a number of processing steps. The result is district-aggregated time series that can readily be used together with the MI data as input to ML algorithms. Depending on the nature of the predictor, only a subset of the processing steps may apply. First, the raw data are converted to a common format (csv). In many cases, the spatial scale is much larger than the region of interest. For instance, the NDVI data are global. The second step is therefore to reduce the data to the region of interest (ROI), namely Augsburg and surroundings.

Point sources, such as station data, are converted to a 1 km grid using a Kriging method. To account for the different resolutions in time interpolation to a common target frequency is conducted. The frequency is based on the highest resolution supported by the MI registry which is daily.

Finally, the data are aggregated to the district level to arrive at a daily time series for each of the predictors. Together with the ground truth, it can be used with all standard ML algorithms for time series prediction. The procedure has been implemented with the Python programming language. Figure 4.5 lists the packages used to carry out the processing steps.

The next step is to apply supervised ML techniques (e.g. Decision Trees, ANN) to the fused data to regress the incidence of MI based on the prepared environmental, socio-economic and climatic predictors. From this, we expect to gain first insights into the importance of heat stressors relative to other risk factors (Marien et al. 2022).

Fig. 4.5 Schematic of the
fusion process chain

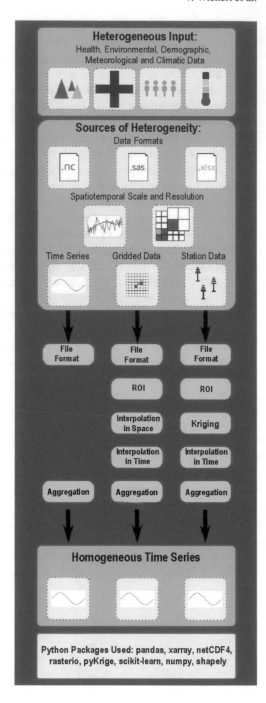

4.4 Point-To-Space Extrapolation

4.4.1 Estimation of Missing Methane Emissions from Offshore Platforms by a Data Analysis of the Emission Database for Global Atmospheric Research (EDGAR)

Disused and active offshore platforms can emit methane, the amount being difficult to quantify. In addition, explorations of the sea floor in the North Sea showed a release of methane near the boreholes of both, oil and gas-producing platforms. The basis of this study is the established Emission Database for Global Atmospheric Research (EDGAR) (Janssens-Maenhout et al. 2019). While methane emission fluxes in the EDGAR inventory and platform locations are matching for most of the oil platforms, almost all of the gas platform sources are missing in the database. We develop a method for estimating the missing sources based on the EDGAR emission inventory.

EDGAR is an inventory from the EC-JRC and Netherlands' Environmental Assessment Agency (Saunois et al. 2016). National reports of greenhouse gas emissions are the basis for emission inventories like EDGAR which is used as emission input for the simulations in this work. It covers sector- and country-specific time series of the period 1970–2012 with monthly resolution and a global spatial resolution of $0.1° \times 0.1°$ providing CH_4, CO_2, CO, SO_2, NO_x, C_2H_6, C_3H_8 and many other species. Different source sectors in EDGAR are defined using the IPCC 1996 guidelines (Janssens-Maenhout et al. 2019). When calculating the sector-specific emissions, a differentiation of emission processes improves and refines the estimates of EDGAR. Therefore, technology-specific emission factors, end-of-pipe abatement measurements, a modelling based on latest scientific knowledge, available global statistics and IPCC-recommended data are used. The emissions are then distributed on maps via proxy datasets based on national spatial data containing information about population density, the road network, waterways, aviation and shipping trajectories (Janssens-Maenhout et al. 2012). A global $0.1° \times 0.1°$ grid is used on which the emissions are assigned to, either as a single-point source (e.g. oil or gas platforms), distributed over a line source (e.g. shiptracks) or over an area source (e.g. agricultural fields) always depending on the source sectors and subsectors. For this work, methane point source emissions from EDGAR are of a high importance. These one-dimensional sources are allocated to a single grid cell of the $0.1° \times 0.1°$ grid with the average of all points that fall into the same cell (Janssens-Maenhout et al. 2019). We are aiming to replace the gridded emissions from EDGAR with point source in our atmospheric model by extrapolating them from point to space to adjust missing emissions within EDGAR and improve the spatial accuracy of the current dataset.

For this study, the global atmospheric model ICON (ICOsahedral Nonhydrostatic model) was used with EDGAR data as input for emissions. ICON is a joint development of the German Weather Service (DWD) and the Max Planck Institute for

Meteorology (MPI-M). Due to its dynamical core, which is based on the nonhydro-static formulation of the vertical momentum equation, simulations with a high horizontal resolution up to grid spacings of a few hundreds of metres are possible. ART (Aerosols and Reactive Trace Gases) is an online-coupled model extension for ICON that includes chemical gases and aerosols. One aim of the model is the simulation of interactions between the trace substances and the state of the atmosphere by coupling the spatiotemporal evolution of tracers with atmospheric processes (Schröter et al. 2018). The point source module in ART takes the prescribed emission fluxes as point sources of substances and adds them to new or existing chemical tracers by distributing the one-dimensional fluxes to the area of the corresponding triangular grid cell in ICON. First of all, the emission factor is calculated through the source strength of the emissions, the area of the grid cell and the model time step as shown in Eq. (1) (Prill et al. 2019, p. 77).

$$ \text{emiss\_fct} = \frac{\text{source\_strength}}{\text{cell\_area}} \cdot \text{dtime} \quad \left[\frac{\text{kg}}{m^2} \right] \tag{1} $$

As a next step, the above-calculated emission factor is added to the actual tracer value of the grid. Equation (2) shows how emiss_fct is multiplied with a height factor. Also the density of air ρ in kg/m^3 and the height of the corresponding ICON layer dz in metres come to play. In our case, all the point sources are on the lowest model level with a height of 20 m.

$$ \text{tracer} = \text{tracer} + \frac{h \cdot \text{emiss\_fct}}{(\rho \cdot \text{dz})} \quad \left[\frac{\text{kg}}{\text{kg}} \right] \tag{2} $$

The ICON-ART sensitivity simulations of this study aimed to investigate the differences between simulations with gridded emissions from EDGAR and point sources of ART. Therefore, the methane emissions in the North Sea Region that are contained in EDGAR were distributed to all 956 point sources representing the offshore platforms in this area. Figure 4.6 displays the procedure of how EDGAR gridded emissions are replaced and adjusted by point sources.

If we compare simulations with gridded EDGAR and simulation with point sources, it is remarkable that both fit quite well over the whole year. This can be seen as a proof of concept that the point source module of ART adjusts the platform emissions of EDGAR in a satisfying way. The maximum of the absolute difference over the year on a global scale is 4.755 ppbv and the difference of the annual mean is − 0.342 ppbv, showing that the point source module slightly overestimates the gridded emissions with a difference of less than 0.02%. These results show that the point source module in ICON-ART can model methane emissions as well as conventional gridded input data with the advantage that the spatial accuracy of this point-to-space method is significantly better.

With the point source module of ICON-ART being a successful tool for point-to-space extrapolation, we are now able to include the missing platforms into the

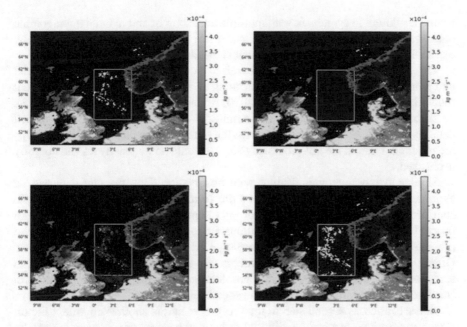

Fig. 4.6 Illustrating the procedure of how point sourceemissions adjust EDGAR gridded emission. The methane emission fluxes of original EDGAR (*upper left*) are removed within the North Sea Region (*white box*) (*upper right*). From all the locations of the platforms (*lower left*), the ones inside the North Sea Region are chosen and the EDGAR emission are equally distributed to them as point sources (*lower right*)

EDGAR dataset and access their influence on the methane distribution on regional (North Sea) and global scales.

4.4.2   Point-to-Space Methods

The aim of this subproject within the Digital Earth project is to develop a data-driven machine learning (ML) and artificial intelligence (AI) method to advance currently available datasets and maps, improve the resolution of observational datasets and interpolate values for locations where no observations are available, infer patterns from simulated data and improve estimations by combining the simulations with measurements.

Data sources at scattered sites (in situ measurements) and gridded data (satellite data, output of numerical models) are combined to construct a higher-resolution map, or predict values at ungauged sites (interpolation, "downscaling"), or to develop simulations for unobserved scenarios—locations and instance of time.

The project is designed as a close interdisciplinary collaboration between data science and Earth scientists. The fundamental concept of a point-to-space problem is to find a solution in agreement with multiple data sources and to build a geographic dataset (or map) based on them. The outcome is a more complete dataset derived by extrapolating into space and time that can be used for further analyses.

This method can be used especially well to extend the sparsely scattered spatial and spatiotemporal maps by the application of ML methods instead of physical modelling approaches (Peng Xie et al. 2020; Amato et al. 2020, Volfová et al. 2012).

To build a procedure which can use multiple covariates from point or gridded datasets as inputs and find an estimator for the outcome variable, the following steps are required. First, in the pre-processing step, the data homogeneity and its ubiquity is checked.

Similar to most methods dealing with observation data, the raw data can be extremely heterogeneous and in various file formats, and adhere to different data standards. They can also differ in spatial and temporal resolution; hence, choosing a proper target grid resolution can be of utmost importance. In this step, all the data are projected into the same format and same target grid since the sources are so diverse. Often the first major issue is to pre-process and also check the observational data for errors and outliers. However, this problem is becoming simpler to solve with standardization of formats and metadata specification.

Normalization as second step in the pre-processing is another important factor, since the covariates can be of a different order, and hence, error propagation can be relevant to their scale (Singh et al. 2020). Determining the proper way to normalize is important, since it will have a direct effect on the algorithm as well as the results. In this step, all the variables, including dependent and independent variables, are normalized and hence prepared for the regression algorithms.

Given the nature of the problem, the respective Earth science expert needs to be consulted especially for the selection of the appropriate covariates. They need to be selected in a way that if the information is not available directly then it can be inferred from another variable or proxy. Moreover, the original point-to-space problem might extend similarly to other point-to-space subproblems for the selected covariates since they might not be available in the higher resolution of the target grid either.

A predictive ML regression model for these subsets of the problem can be trained and used to predict each of the covariates. For each of these covariates, a different ML method can be applied based on its characteristics and physics of the phenomena/covariate and the required precision. Consequently, after solving these subsets, the selected covariates are approximated on the target grid. The trained ML's prediction method can be applied, and this means that all the necessary requirements for the final step are now fulfilled.

In the final step, a more sophisticated ML regression method is applied for the target variable by combining the previously estimated covariates. Here, methods such as co-kriging to find optimal solutions for the outcome variable and covariates simultaneously can prove useful.

Similar to all ML methods, data splitting methods are employed and a percentage of the data is chosen as training data, with the remaining data used for testing and validation.

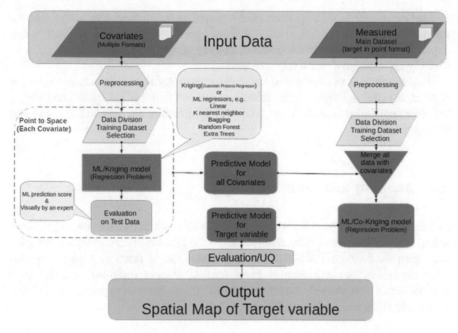

Fig. 4.7 Process chain schematic

Shown in Fig. 4.7 is a simplistic schematic of the suggested procedure. Any typical regressor can be easily implemented and used here, even a modified kriging method. Examples of the implemented and evaluated regression methods are linear, K-nearest neighbours, bagging trees, extra trees, random forest and simple ordinary kriging. Tree methods have been shown to have the advantage of high accuracy of the prediction for interpolations in the point-to-space problem. (Wessel et al. 2018).

To apply the same procedure to other similar problems, the following considerations need to be contemplated.

As with all ML regression problems, a regressor can lead to overfitting and some mitigation needs to be considered. One way is to employ different data splitting schemes and to check the number of features selected.

In co-kriging, similar to the normal kriging method, a kernel function is used based on a covariance matrix. Co-kriging is a conditional random field generator which superposes a method based on kriging with a multivariate Gaussian method building on a covariance matrix (Volfová et al. 2012). Determining the covariance function for co-kriging is not easy, since the methods with dependent covariates are so sensitive to the input and the stability of the system depends on the condition number of the prior (Putter et al. 2001; Ababou et al. 1994). Applying a co-kriging method enables a flexible feature advantage and makes it trivial to utilize any other data or information on new covariates/proxies which are available later, and as a result, the accuracy of the method can always be improved in this way.

Finally, uncertainty in data and methods should be evaluated to ensure that the method is based on a proper stable solution. Again, the evaluation of results using expert knowledge requires deep understanding of the problem, since data can be visually appealing but extremely erroneous.

This research is an example of a successful new collaboration between multiple centres involved in Digital Earth. It has been shown that it is indispensable to bring together knowledge from different fields and applications for developing successful applications.

4.5 Anomaly and Event Detection

In the Digital Earth Flood Showcase, we strive to understand hydrological extreme events across disciplines, from river basins to the sea. Comprehending the complex and often time-delayed chain of events surrounding a flood is a compartment-spanning task in the Earth sciences. Here, we feature an example from the showcase that illustrates how anomaly detection and investigation can be supported by suitable data exploration and analysis methods.

4.5.1 *Computational Methods for Investigating the Impacts of the Elbe Flood 2013 on the German Bight*

At the end of a hydrological extreme event chain, river water is discharged into the ocean, transporting matter and thereby, unusual amounts of nutrients and pollutants into the coastal system. To investigate the impacts of riverine floods on the marine environment, several steps need to be undertaken: the river plume needs to be detected and its spatial extent and development in time need to be determined to identify the study region and time interval. The procedure relies on the combination of heterogeneous data sources, such as in situ measurements, matching model data and additional satellite data for a wider spatial coverage. This article features three computational methods that are crucial for successfully identifying and investigating the processes in and around a river plume after a riverine flood event: automatic anomaly detection, producing customized model trajectories and generating time series of productivity. The complete workflow will be described in more detail in Sect. 5.3.3. under "The River Plume Workflow".

Automatic Anomaly Detection

An enhanced feature of the River Plume Workflow is the automatic detection of parameter anomalies in both in situ FerryBox measurements during summer operations and year-round satellite data.

Three separate definitions of anomalies caused by a flood event originating from a freshwater river mouth are considered. They include: increased chlorophyll levels;

decreased salinity; and, particularly during the winter where the river water may be cooler than the sea, a decrease in sea surface temperature.

As an alternative to manually searching for anomalies in the River Plume Workflow's interactive map (see Sect. 3.2), users can select years of interest to undergo a Gaussian regression-based statistical analysis (Pedregosa et al. 2011), which provides the user with a list of recommended dates of interest. Gaussian processes (GPs) are powerful and flexible models for modelling time series data, which makes them a practical option for anomaly detection. Here, a Gaussian regression analysis is performed to generate a posterior probability distribution based on the daily parameters over the selected time period. This method requires a specified prior distribution, with the prior's covariance given by a kernel, in this case, a generalized Matern kernel with an amplitude factor and an observation noise component. The model uses high smoothing capabilities for more efficient anomaly detection. Also generated are posterior standard deviations that create a 95% confidence region around the posterior distribution. Anomalies are recorded when the measured data and its associated uncertainties fall completely outside the confidence region, either above or below depending on which parameter is being considered. Figure 4.8 gives an example of the outputted results for chlorophyll during 2013.

Future development would involve near real-time automatization to detect anomalies as they occur based on previous established annual patterns. This would involve taking an average of the posterior distributions over a number of years and detecting an anomaly in near real-time data if it deviates from this average.

Produce Customized Model Trajectories

Fig. 4.8 Daily chlorophyll measurements with one region of interest (RoI) for in situ FerryBox (*upper*) and satellite (*lower*) data. The continuous line represents the Gaussian regression generated distribution, while the *shaded grey* band is the confidence region. The *red circles* indicate detected anomalies

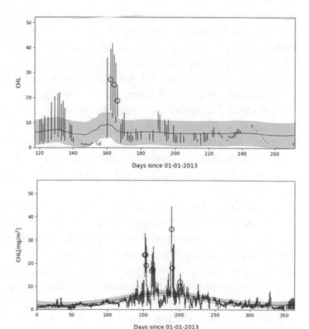

A crucial component of the River Plume Workflow is the computation of model trajectories for observational FerryBox data to not only detect the river plume, but also determine its spatial and temporal extent in the ocean.

FerryBox transects provide, often regular, observational data on the study region. The data are made available in near real time via the COSYNA (Coastal Observing System for Northern and Arctic Seas) data portal (Baschek et al. 2017; Breitbach et al. 2016). The PELETS-2D code (Callies et al. 2021) is used to compute model trajectories for each FerryBox transect. These model trajectories consist of the measured water bodies' positions up to ten days before and after the actual measurement as simulated by the numeric model. The simulated positions are then recombined into synoptic maps, featuring time shifted positions of all observations at one point in time. While the model trajectories are useful to investigate an anomaly's origin, the synoptic maps show the spatial and temporal extent of the river plume on its presumed path across the ocean.

The simulations and their combination in synoptic plots are too computationally expensive to be done in real time. Therefore, they are currently only produced for specific events. However, work is currently underway to optimize the code for operational use and make the model data available in near real time.

Generate Time Series of Productivity

Combining observational data with specifically produced model trajectories allows users to determine the spatial and temporal extent of a river plume in the ocean, but does not help with understanding the processes that happen inside these waterbodies. For that reason, another method of the River Plume Workflow focuses on blending the spatial information of the model trajectories with the parameter information from satellite data. We use integrated satellite datasets taken from the Copernicus Marine Environment Monitoring System (CMEMS 2021). The datasets give daily average chlorophyll, salinity and sea surface temperature measurements after full processing, which includes the reconstruction of cloud-covered areas. The chlorophyll datasets have a 1 km^2 tiled resolution, while salinity and sea surface temperature have a 7×11.6 km^2 tiled resolution.

Our method automatically extracts parameter values from satellite data for the locations on a selected waterbodies' modelled trajectory, thus producing a time series of values along the modelled pathway of the waterbody. As an example, this method enables researchers to produce chlorophyll time series for waterbodies associated with the river plume and therefore to investigate chlorophyll degradation rates inside the river plume.

The methods described here were implemented into the River Plume Workflow, a scientific workflow prototype (see Sect. 5.3.3), and tested for the Elbe flood event from 2013. A ferry equipped with a FerryBox regularly covers the Büsum-Heligoland line during the period from April to October. These transects are highly relevant for our example as they cover a region close to the Elbe outflow. Using the anomaly detection algorithm, several instances where the Elbe river plume potentially crossed the FerryBox transect were determined, such as on 23 June 2013, where an anomaly in salinity and temperature was visible (see Fig. 4.9). Simulated trajectories of the relevant water bodies across the North Sea point to the Elbe River as the anomaly's

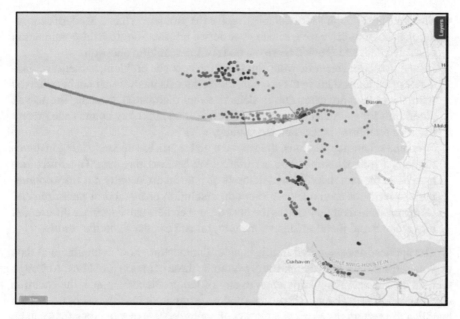

Fig. 4.9 Screenshot of the River Plume Workflow's interactive map. The selected region marks the area of the suspected river plume on the FerryBox transect on 23 June 2013. *Blue dots* highlight one water body's modelled trajectory originating in the Elbe River

origin. The generated time series of chlorophyll gives an overview of productivity changes in the region during and after the flood event. This helped identify promising study regions in the North Sea regarding noteworthy biological events, e.g. unusual algae blooms after a riverine flood event. In general, the River Plume Workflow helped to give better understanding of the sequence of events of the Elbe river flood of 2013 and its impacts on the marine environment of the North Sea.

For more information on the River Plume Workflow, please see https://digitalea rth-hgf.de/results/workflows/flood-event-explorer/#accordion-4.

4.6 Conclusions

All examples provided in this chapter have in common that they apply advanced computational approaches to gain new insights from existing data and help to further our process understanding through innovative data analysis. Regardless of the scientific question or exact context, the presented computational approaches address problems that occur frequently in modern Earth and Environmental sciences.

They typically fall into one or more of the following groups:

- There are not enough data available to solve the problem at hand. Modern computational methods can enable scientists to derive information from the combination of related data and provide more context for the scientific question.
- In contrast, some scientific workflows face the opposite problem, namely an abundance of available data. In these cases, insightful data analysis cannot be achieved with classical approaches due to data size and distributed storage. The use of algorithms for automatically classifying events and providing context can greatly improve scientists' process understanding.
- For some scientific questions, the way to a better process understanding involves the comparison or combination of different kinds of existing data. The difficulties here lie in the fact that different datasets of interest are usually not fully comparable in terms of spatial and/or temporal resolution or method of measurement. Computational methods such as the ones described here can take these differences into account and therefore ensure meaningful and correct scientific results.

The applications presented in this chapter demonstrate how computational data exploration and data analytics can help overcome these common problems. Although there is no one-size-fits-all approach to any of the problems we face, the example applications in this chapter show that modern computational approaches can help handling the current paradoxical situation of having access to more data than can be handled by classical methods, while simultaneously needing to overcome a lack of data in other contexts. Moreover, the examples show that those approaches enable us to create additional benefit from the available data and improve our understanding of the complex system Earth.

Acknowledgements The KORA study was initiated and financed by the Helmholtz Zentrum München—German Research Center for Environmental Health, which is funded by the German Federal Ministry of Education and Research (BMBF) and by the State of Bavaria. Furthermore, KORA research was supported within the Munich Center of Health Sciences (MC-Health), Ludwig-Maximilians-Universität, as part of LMUinnovativ. Since 2000, the MI data collection has been co-financed by the German Federal Ministry of Health and Social Security to provide population-based MI morbidity data for the official German Health Report (see www.gbe-bund.de).

References

Ababou R, Bagtzoglou AC, Wood EF (1994) On the condition number of covariance matrices in kriging, estimation, and simulation of random fields. Math Geol 26:99–133. https://doi.org/10.1007/BF02065878

Abadi M, Agarwal A, Barham P, Brevdo E, Chen Z, Citro C, Corrado GS, Davis A, Dean J, Devin M, Ghemawat S, Goodfellow I, Harp A, Irving G, Isard M, Jia Y, Jozefowicz R, Kaiser L, Kudlur M, Levenberg J, Mane D, Monga R, Moore S, Murray D, Olah C, Schuster M, Shlens J, Steiner B, Sutskever I, Talwar K, Tucker P, Vanhoucke V, Vasudevan V, Viegas F, Vinyals O, Warden P, Wattenberg M, Wicke M, Yu Y, Zheng X (2016) Tensorflow: large-scale machine learning on heterogeneous distributed systems

Amato F, Guignard F, Robert S et al (2020) A novel framework for spatio-temporal prediction of environmental data using deep learning. Sci Rep 10:22243. https://doi.org/10.1038/s41598-020-79148-7

Appel KW, Pouliot GA, Simon H, Sarwar G, Pye HOT, Napelenok SL, Akhtar F, Roselle SJ (2013) Evaluation of dust and trace metal estimates from the Community Multiscale Air Quality (CMAQ) model version 5.0. Geosci Model Dev 6:883–899. https://doi.org/10.5194/gmd-6-883-2013

Baschek B, Schroeder F, Brix H, Riethmüller R, Badewien TH, Breitbach G, Brügge B, Colijn F, Doerffer R, Eschenbach C, Frie-drich J, Fischer P, Garthe S, Horstmann J, Krasemann H, Metfies K, Merckelbach L, Ohle N, Petersen W, Pröfrock D, Röttgers R, Schlüter M, Schulz J, Schulz-Stellenfleth J, Stanev E, Staneva J, Winter C, Wirtz K, Wollschläger J, Zielinski O, Ziemer F (2017) The coastal observing system for northern and arctic seas (COSYNA). Ocean Sci 13:379–410. https://doi.org/10.5194/os-13-379-2017

Bieser J, Aulinger A, Matthias V, Quante M, Builtjes P (2011) SMOKE for Europe—adaptation, modification and evaluation of a comprehensive emission model for Europe. Geosci Model Dev 4:47–68. https://doi.org/10.5194/gmd-4-47-2011

BLFS: Bayerisches Landesamt für Statistik: GENESIS Datenbank. https://www.statistikdaten.bay ern.de/genesis/online/. Last Accessed on 01 September 2021

BLFU: Bayerische Landesamt für Umwelt: Lufthygienische Landesüberwachungssystem Bayern (LÜB). https://www.lfu.bayern.de/luft/immissionsmessungen/messwertarchiv/index.htm. Last Accessed on 01 September 2021

Breitbach G, Krasemann H, Behr D, Beringer S, Lange U, Vo N, Schroeder F (2016) Accessing diverse data comprehensively—CODM, the COSYNA data portal. Ocean Sci 12:909–923. https://doi.org/10.5194/os-12-909-2016

Callies U, Kreus M, Petersen W, Voynova YG (2021) On using Lagrangian drift simulations to aid interpretation of in situ monitoring data. Front Mar Sci 8:769. https://doi.org/10.3389/fmars.2021.666653

Chen K, Breitner S, Wolf K, Hampel R, Meisinger C, Heier M, Von Scheidt W, Kuch B, Peters A, Schneider A (2019) Temporal variations in the triggering of myocardial infarction by air temperature in Augsburg, Germany, 1987–2014. Eur Heart J 40:1600–2160. https://doi.org/10.1093/eurheartj/ehz116

CMEMS North Atlantic Chlorophyll (Copernicus-GlobColour) from Satellite Observations. Daily Interpolated (Reprocessed from 1997). Copernicus Monitoring Environment Marine Service (CMEMS). Available at https://resources.marine.copernicus.eu/product-detail/OCEANC OLOUR_ATL_CHL_L4_REP_OBSERVATIONS_009_098/. Accessed 21 September 2021

CMEMS Atlantic-European North West Shelf-Ocean Physics Reanalysis. Copernicus Monitoring Environment Marine Service (CMEMS). Available at https://resources.marine.copernicus.eu/pro duct-detail/NWSHELF_MULTIYEAR_. Accessed 21 September 2021

Franco B, Mahieu E, Emmons LK, Tzompa-Sosa ZA, Fischer EV, Sudo K, Bovy B, Conway S, Griffin D, Hannigan JW, Strong K, Walker KA (2016) Evaluating ethane and methane emissions associated with the development of oil and natural gas extraction in North America. Environ Res Lett 11:044010. https://doi.org/10.1088/1748-9326/11/4/044010

Greinert J, Schoening T (n.d.) RV POSEIDON Fahrtbericht/Cruise Report POS526—SeASOM: Semi-Autonomous Subsurface Optical Monitoring for methane seepage and cold-water coral studies in the North Sea, Bergen (Norway)—Dogger Bank (Netherlands)—Hirtshals (Denmark)—Tisler (Norway)—[WWW Document]. Report. https://doi.org/10.3289/geomar_rep_ns_51_2019

Holle R, Happich M, Löwel H, Wichmann HE (2005) KORA—A research platform for population based health research. Gesundheitswesen 67:19–25. https://www.doi.org/10.1055/s-2005-858235

Jacob D, Petersen J, Eggert B, Alias A, Bossing Christensen O, Bouwer LM, Braun A, Colette A, Deque M, Georgievski G, Georgopoulou E, Gobiet A, Menut L, Nikulin G, Haensler A, Hempelmann N, Jones C, Keuler K, Kovats S, Kröner N, Kotlarski S, Kriegsmann A, Martin E, Van Meijgaard E, Moseley C, Pfeifer S, Preuschmann S, Rademacher C, Radtke K, Rechid

D, Rounsevell M, Samuelsson P, Somot S, Soussana JF, Teichmann C, Valentini R, Vautard R, Weber B, Yiou P (2014) EURO-CORDEX: new high-resolution climate change projections for European impact research. Reg Environ Change 14(2):563–578. https://doi.org/10.1007/s10113-013-0499-2

Janssens-Maenhout G, Crippa M, Guizzardi D, Muntean M, Schaaf E, Dentener F, Bergamaschi P, Pagliari V, Olivier JGJ, Peters JAHW, van Aardenne JA, Monni S, Doering U, Petrescu AMR, Solazzo E, Oreggioni GD (2019) EDGAR v4.3.2 global atlas of the three major greenhouse gas emissions for the period 1970–2012. Earth Syst Sci Data 11:959–1002. https://doi.org/10.5194/essd-11-959-2019

Janssens-Maenhout G, Pagliari V, Guizzardi D, Muntean M (2012) Global emission inventories in the Emission Database for Global Atmospheric Research (EDGAR) Manual (I) Gridding: EDGAR emissions distribution on global gridmaps. European Commission—Joint Research Centre—Institute for Environment and Sustainability

Marien L, Valizadeh M, Zu Castell W, Nam C, Rechid D, Schneider A, Meisinger C, Linseisen J, Wolf K, Bouwer LM (2022) Machine learning models to predict myocardial infarctions from past climatic and environmental conditions. Nat Haz Earth Syst Sci Disc. https://doi.org/10.5194/nhess-2021-389

NRW. Open Geo Data, State of North-Rhine Westphalia. https://www.opengeodata.nrw.de/produkte/. Last Aaccessed on 01 September 2021

Pedregosa F, Varoquaux G, Gramfort A, Michel V, Thirion B, Grisel O et al (2011) Scikit-learn: machine learning in python. J Mach Learn Res 12, 2825–2830

Prill F, Reiner D, Rieger D, Zängl G, Schröter J, Förstner J, Werchner S, Weimer M, Ruhnke R, Vogel B (2019) ICON model tutorial. Working with the ICON model, practical exercises for NWP mode and ICON-ART. Deutscher Wetterdienst, Karlsruhe Institute of Technology, Max-Planck-Institut für Meteorologie

Putter H, Young GA (2001) On the effect of covariance function estimation on the accuracy of kriging predictors. Bernoulli 7(3):421–438. https://projecteuclid.org/euclid.bj/1080004758

Ronneberger O, Fischer P, Brox T (2015) U-net: convolutional networks for biomedical image segmentation. Medical Image Computing and Computer-Assisted Intervention (MICCAI), Springer, LNCS, Vol. 9351: 234–241. Available at arXiv:1505.04597

Saunois M, Bousquet P, Poulter B, Peregon A, Ciais P, Canadell JG, Dlugokencky EJ, Etiope G, Bastviken D, Houweling S, Janssens-Maenhout G, Tubiello FN, Castaldi S, Jackson RB, Alexe M, Arora VK, Beerling DJ, Bergamaschi P, Blake DR, Brailsford G, Brovkin V, Bruhwiler L, Crevoisier C, Crill P, Covey K, Curry C, Frankenberg C, Gedney N, Höglund-Isaksson L, Ishizawa M, Ito A, Joos F, Kim H-S, Kleinen T, Krummel P, Lamarque J-F, Langenfelds R, Locatelli R, Machida T, Maksyutov S, McDonald KC, Marshall J, Melton JR, Morino I, Naik V, O'Doherty S, Parmentier F-JW, Patra PK, Peng C, Peng S, Peters GP, Pison I, Prigent C, Prinn R, Ramonet M, Riley WJ, Saito M, Santini M, Schroeder R, Simpson IJ, Spahni R, Steele P, Takizawa A, Thornton BF, Tian H, Tohjima Y, Viovy N, Voulgarakis A, van Weele M, van der Werf GR, Weiss R, Wiedinmyer C, Wilton DJ, Wiltshire A, Worthy D, Wunch D, Xu X, Yoshida Y, Zhang B, Zhang Z, Zhu Q (2016) The global methane budget 2000–2012. Earth Syst Sci Data 8:697–751. https://doi.org/10.5194/essd-8-697-2016

Schröter J, Rieger D, Stassen C, Vogel H, Weimer M, Werchner S, Förstner J, Prill F, Reinert D, Zängl G, Giorgetta M, Ruhnke R, Vogel B, Braesicke P (2018) ICON-ART 2.1: a flexible tracer framework and its application for composition studies in numerical weather forecasting and climate simulations. Geosci Model Dev 11:4043–4068. https://doi.org/10.5194/gmd-11-4043-2018

Sieck K, Nam C, Bouwer LM, Rechid D, Jacob D (2020) Weather extremes over Europe under 1.5 and 2.0 °C global warming from HAPPI regional climate ensemble simulations. Earth Syst Dyn 12(2):457–468. https://doi.org/10.5194/esd-2020-4

Singh D, Singh B (2020) Investigating the impact of data normalization on classification performance. Appl Soft Comput 97(Part B):105524. ISSN 1568-4946. https://doi.org/10.1016/j.asoc.2019.105524

Van Dingenen R, Crippa M, Anssens-Maenhout G, Guizzardi D, Dentener F (2018) Global trends of methane emissions and their impacts on ozone concentrations. JRC Sci Policy Rep. https://doi.org/10.2760/8201755

Visschedijk AJH, Denier Van Der Gon HAC, Doornenbal HC, Cremonese L (2018) Methane and ethane emission scenarios for potential shale gas production in Europe. Adv Geosci 45:125–131. https://doi.org/10.5194/adgeo-45-125-2018

Vlasenko A, Matthias V, Callies U (2021) Simulation of chemical transport model estimates by means of neural network using meteorological data. Atmos Environ. https://doi.org/10.1016/j.atmosenv.2021.118236

Volfová A, Šmejkal M (2012) Geostatistical methods in R. Geoinformatics FCE CTU 8:29–54. https://doi.org/10.14311/gi.8.3

Wagenaar D, De Jong J, Bouwer LM (2017) Multi-variable flood damage modelling with limited data using supervised learning approaches. NHESS 17(9):1683–1696. https://doi.org/10.5194/nhess-17-1559-2017

Wessel M, Brandmeier M, Tiede D (2018) Evaluation of different machine learning algorithms for scalable classification of tree types and tree species based on sentinel-2 data. Remote Sens 10:1419. https://doi.org/10.3390/rs10091419

Xie P, Li T, Liu J, Du S, Yang X, Zhang J (2020) Urban flow prediction from spatiotemporal data using machine learning: a survey. Inf Fusion 59: 1–12. ISSN 1566–2535. https://doi.org/10.1016/j.inffus.2020.01.002

Yacovitch TI, Daube C, Herndon SC (2020) Methane emissions from offshore oil and gas platforms in the Gulf of Mexico. Environ Sci Technol 54:3530–3538. https://doi.org/10.1021/acs.est.9b07148

Chapter 5
Data Analysis and Exploration with Scientific Workflows

Doris Dransch, Daniel Eggert, Nicola Abraham, Laurens M. Bouwer, Holger Brix, Ulrich Callies, Thomas Kalbacher, Stefan Lüdtke, Bruno Merz, Christine Nam, Erik Nixdorf, Daniela Rabe, Diana Rechid, Kai Schröter, Bente Tiedje, Dadiyorto Wendi, and Viktoria Wichert

Abstract Geoscientific data analysis has to face some challenges regarding seamless data analysis chains, reuse of methods and tools, interdisciplinary approaches and digitalization. Computer science and data science offer concepts to face these challenges. We took the concepts of scientific workflows and component-based software engineering and adapted it to the field of geoscience. In close collaboration of computer and geo-experts, we set up an expedient approach and technology to develop and implement scientific workflows on a conceptual and digital level. We applied the approach in the showcase "Cross-disciplinary Investigation of Flood Events" to introduce and prove the concepts in our geoscientific work environment, and assess how the approach tackles the posed challenges. This is exemplarily demonstrated with the Flood Event Explorer which has been developed in Digital Earth.

Keywords Workflow · Digitalization · Component-based software · Software engineering · Reusability · Data Science

5.1 Challenges and Needs

The digitalization of science offers computer science and data science approaches that can improve scientific data analysis and exploration from several perspectives. Digital

D. Dransch (✉) · D. Eggert · S. Lüdtke · B. Merz · D. Rabe · K. Schröter · D. Wendi
Helmholtz Centre Potsdam - GFZ German Research Centre for Geosciences, Potsdam, Germany
e-mail: doris.dransch@gfz-potsdam.de

N. Abraham · H. Brix · U. Callies · V. Wichert
Helmholtz-Zentrum Hereon, Geesthacht, Germany

L. M. Bouwer · C. Nam · D. Rechid · B. Tiedje
Climate Service Center Germany (GERICS), Helmholtz-Zentrum Hereon, Hamburg, Germany

T. Kalbacher · E. Nixdorf
Helmholtz Centre for Environmental Research - UFZ, Leipzig, Germany

© The Author(s) 2022
L. M. Bouwer et al. (eds.), *Integrating Data Science and Earth Science*,
SpringerBriefs in Earth System Sciences,
https://doi.org/10.1007/978-3-030-99546-1_5

Earth applies and adapts the concepts of scientific workflows and component-based software engineering to address the following challenges and needs:

1. Scientific data analysis and exploration can be seen as a process where scientists fulfil several analytical tasks with a variety of methods and tools. Currently, scientific data analysis is often characterized by performing the analytical tasks in single isolated steps with several isolated tools. This requires many efforts for scientists to bring data from one tool to the others, to integrate and analyse data from several sources and to combine several analysis methods. This isolated work environment hinders scientists to extensively exploit and analyse the available data. We need enhanced work environments that integrate methods and tools into seamless data analysis chains and that allow scientists to comprehensively analyse and explore spatio-temporal, multivariate datasets from various sources that are common in geoscience.

2. Scientific data analysis and exploration often requires specific, highly tailored methods and tools; many of them are developed by geoscientists themselves. Often the methods and tools can hardly be shared since they miss state-of-the-art concepts and techniques from computer science. In consequence, the analysis methods and tools are not available for others and have to be invented again and again. This costs time and money that cannot be spent for scientific discovery. Therefore, a further requirement for advanced scientific data analysis is to facilitate the sharing and reuse of specific analytical methods and tools.

3. Scientific data analysis and exploration is more and more embedded in an interdisciplinary context. To answer complex questions relevant to society, such as concerning drivers and consequences of global change, sustainable use of resources, or causes and impacts of natural hazards, needs to integrate knowledge from different scientific communities. Data from various sources have to be integrated, but also the data analysis approaches itself that extract information from the data have to be linked across communities. This is similar to the concept of coupling physics-based models which is a well-established method in geoscience to investigate and understand related processes in the Earth system. Integrating the analytical approaches across disciplines requires efforts at two levels: integration on the technical executable level, but also integration on the conceptual scientific level. On the one hand, we need technical environments that facilitate the integration of methods and tools; on the other hand, we need suitable means to support the exchange of scientific knowledge. Means are required that make apparent the data analysis approaches of other scientific communities, their scientific objective, the data that are needed as input, and the output that is generated, the methods that are applied, and the results that are created.

4. The transformation of science into digital science has been an ongoing process for many years. To get the best possible results out of the process, a close collaboration of geo- and computer experts is needed. Geoscientists have to clarify and communicate comprehensibly their scientific needs, and computer and data scientists have to understand these requirements and transform them

into their own approaches and solutions. Suitable means are required to facilitate this transformation. This is especially true since both disciplines have rather different working concepts: geoscientists work mostly application-oriented, and computer scientists work on a more generic, formal and abstract basis.

The concepts of scientific workflows and component-based software engineering provide a suitable frame to tackle these needs. Workflows have been applied in science for more than a decade; today, they extend to data-intensive workflows exploiting diverse data sources in distributed computing platforms (Atkinson 2017). Component-based software engineering is a reuse-based approach to defining, implementing and composing loosely coupled independent software components into larger software environments such as scientific workflows (McIlroy 1969; Heineman et al. 2001). We applied the concepts in a showcase, the cross-disciplinary investigation of flood events to introduce and prove the concepts in our geoscientific work environment, and assess how the approaches can tackle the addressed challenges.

5.2   Scientific Workflows

5.2.1   The Concept of Scientific Workflows

The concept of workflow was formally defined by the Workflow Management Coalition (WfMC) as "the computerized facilitation or automation of a business process, in whole or part" (Hollingsworth 1994). Workflows consist of a series of activities with input and output data and are directed to a certain objective. Originally geared towards the description of business processes, workflows have been increasingly used to describe scientific experiments and data analysis processes (Cerezo et al 2013). At the beginning, scientific workflows were focused on authoring and adapting processing tasks to distributed high performance computing. Today, they extend to data-intensive workflows exploiting rich and diverse data sources in distributed computing platforms (Atkinson et al 2017).

Scientific workflows provide a systematic way of describing data analysis with its analytical activities, methods and data needed. Cerezo et al (2013) distinguish three abstraction levels of scientific workflow descriptions: the conceptual, abstract and concrete levels (see Fig. 5.1). On the conceptual level, a workflow is described as a series of activities with input and output data in the language and concepts of the scientist; on the abstract level, the conceptual workflow is mapped to methods and tools to execute the activities; the concrete level provides an executable workflow within a concrete IT infrastructure.

The description of workflows on different abstraction levels provides a number of benefits. It enables scientists to document and communicate scientific approaches and knowledge creation processes in a structured way. It provides the interface between scientific approaches and computing infrastructures. And it allows for sharing, reuse and discovery of workflows or parts of it (Cerezo et al. 2013).

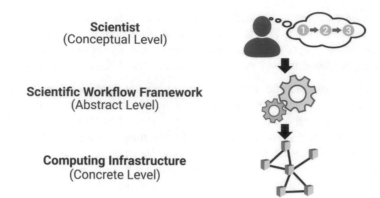

Fig. 5.1 Three abstraction levels of scientific workflow descriptions adapted from Cerezo et al. (2013)

A recent review paper about past, presence and future of scientific workflows points out the strength and needs of scientific data-intensive workflows: "With the dramatic increase of primary data volumes and diversity in every domain, workflows play an ever more significant role, enabling researchers to formulate processing and analysis methods to extract latent information from multiple data sources and to exploit a very broad range of data and computational platforms" (Atkinson et al. 2017, p. 2016).

Several scientific workflow systems have been developed so far to enable scientists making use of the mentioned advantages. Examples are Galaxy, Kepler, Taverna and Pegasus [Workflow Systems 2021].

5.2.2 Scientific Workflows in Digital Earth

We predominantly applied the concept of workflows to structure and describe scientific data analysis approaches in a systematic way and to implement seamless data analysis chains as executable digital workflows. Our objective was not to develop or introduce big workflow engines to model and automatically create scientific workflows. Our focus was the elicitation and description of conceptual workflows and the development of digital workflows facilitating the integration of any data analysis method and tool.

We described and implemented scientific workflows according to the three abstraction levels shown in Fig. 5.1. We slightly adapted the second level to our needs. Since we do not apply workflow engines but manually transform and implement scientific workflows, we consider it as a digital implementation level. To implement the digital workflows, we developed the component-based Data Analytics Software Framework (DASF) (chapter 5.2.3). It facilitates the definition of reusable software components and integrates them into seamless digital workflows.

Our process of developing workflows covered the following steps. First, we created *conceptual workflows*. Several approaches exist to model and describe conceptual workflows such as flowcharts, Petri-Nets, BPMN- or UML-Diagrams (BPMN = Business Process Model and Notation, UML= Unified Modeling Language). We decided on flowcharts since geoscientists are familiar with this straightforward type of presentation. The conceptual workflows are clearly structured records of scientific approaches and knowledge. We used them for communication and discussion between geoscientists, and to figure out how data analysis approaches can be integrated across disciplines. Conceptual workflow descriptions also served as communication means to bridge approaches from geo- and computer science. To elicit the conceptual workflows, we conducted a structured task analysis to determine the scientists' goals, the analytical tasks (which are the activities in a workflow) and the input/output data. Structured task analysis is a well-established approach in requirement analysis (Jonassen et al 1998; Schraagen et al 2000). Interviews, record keeping and activity sampling are methods to gather the required information; tables and diagrams are means to document the results. Figure 5.2 shows exemplarily the River Plume Workflow as one of the conceptual workflows we have modelled and implemented for the Digital Earth Flood Event Explorer (Sect. 5.3).

In the next step, we transformed the conceptual workflow into a *digital workflow*. First, we mapped each analytical task of the conceptual workflow to a suitable method that fulfils the task. We documented this in a mapping list. Table 5.1 shows the mapped

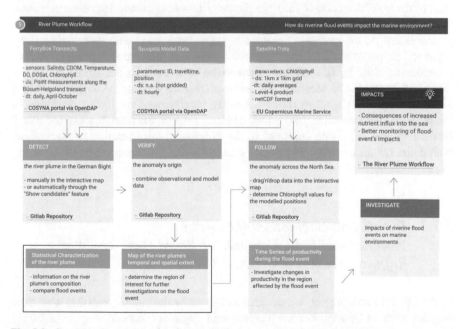

Fig. 5.2 Conceptual workflow description: It documents objective (red), analytical tasks (green), input/output data (blue) of a workflow, in this case the River Plume Workflow of the Digital Earth Flood Event Explorer

Table 5.1 Task to methods mapping list

Step	Subtask	Method	Input Data	Output (Data)
Task: DETECT				
1	Identify regions with low salinity	Visualize observation data with colour-coded salinity on the background map	- Reference date - FerryBox transects	Visualization of water body measurements
2	Confirm riverine origin of water body measurements	Visualize forwards and backwards projected data on the background map	- Time range - Synoptic (projected) data	Visualization of projected water body measurements
3a	Manually define the river plume extent as a polygon	Define the river plume extent by clicking a polygon on the map	- Visualization of projected water body measurements - User input	River plume polygon
3b	Automatically detect the river plume	Trigger automatic anomaly detection via button	- Observation data	River plume polygon and reference date

steps for the "DETECT" task of the River Plume Workflow. In order to address the goal of the task "detection of the river plume", it is broken down into three subtasks. Each of the subtasks is mapped to a certain method, e.g. "Manually define the river plume extent" is mapped to "clicking a polygon on a map" with its corresponding input and output data. All methods defined by this mapping are then implemented as components of the digital workflow based on the Data Analytics Software Framework (DASF) (chapter 5.2.3). Methods that are implemented with the DASF can be shared and reused in other digital workflows. Finally, we *deployed the digital workflows* to the available infrastructure. In our Digital Earth project, the workflows of the Flood Event Explorer have been deployed to the distributed infrastructure of the various Helmholtz research centres. Figure 5.3 shows the concrete deployment of the River Plume Workflow.

5.2.3 Digital Implementation of Scientific Workflows with the Component-Based Data Analytics Software Framework (DASF)

In the Digital Earth project, we decided not to apply the already existing workflow engines such as Galaxy, Kepler, Taverna and Pegasus [Workflow Systems 2021] to model and implement scientific workflows. Although these workflow systems are powerful tools, they have some shortcomings. The intellectual hurdles to be mastered when dealing with workflow systems are high and the systems often do not

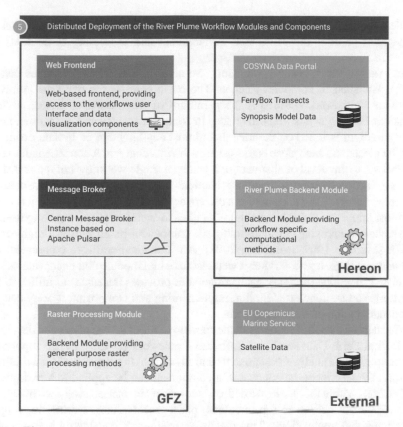

Fig. 5.3 Distributed deployment of the River Plume Workflow modules and components

offer the flexibility that is necessary for the highly exploratory data analysis workflows we have in science (Atkinson et al 2017). For that reason, we decided to follow another approach that is less complex and of higher flexibility. We developed the Data Analytics Software Framework (DASF) to manually implement digital workflows. DASF facilitates the integration of data analysis methods into seamless data analysis chains on the basis of the component-based software engineering paradigm and it enables extensive data exploration by interactive visualization. The integration of data analysis methods into seamless data analysis chains on the basis of the component-based software engineering is described in this chapter. The extensive data exploration by interactive visualization is presented in chapter 3.2 in this book.

The Digital Earth "Data Analytics Software Framework DASF" is targeted to specific requirements of scientific data analysis workflows derived from the challenges formulated in chapter 5. This is the ability to (a) deal with highly explorative scientific workflows, (b) define single data analysis methods and combine them into digital workflows, (c) reuse single methods and workflows or parts of it, (d) integrate already existing and established methods, tools and data into the workflows,

(e) deal with heterogeneous software development and execution environments and (f) support parallel and distributed development and processing of methods and software.

In order to meet these requirements, we combined several approaches that are well established in computer science. The basic concept for our Data Analytics Software Framework (DASF) is the paradigm of "separation of concerns, SoC" (Dijkstra 1982) which is well established in computer science and many other fields. This paradigm is based on the idea that almost anything can be broken down into smaller pieces and each piece is addressing a distinct concern. A concern in this scope can belong to any level of abstraction; it can be a single workflow component (from web services and resources to atomic functions), a workflow or also a more complex linked workflow. Since this paradigm can cover any degree of complexity, it is one of the most important and fundamental principles in sustainable software engineering. One technique applying the SoC paradigm is component-based software engineering, CBSE (McIlroy 1969; Heineman 2001). An individual software component is a software package, a web service, a web resource or a module that encapsulates a set of related functions (or data); each component provides interfaces to utilize it. It is a reuse-based approach to defining, implementing and composing loosely coupled independent components into systems.

We adapted the CBSE concept to the specific needs of data analysis and developed the Digital Earth Data Analytics Software Framework (DASF). DASF supports to define single data analysis components and to connect them to scientific workflows. As initially described, the concept of a component can be applied to any degree of abstraction. Within the DASF, we define six levels of abstraction, as shown in Fig. 5.4. They are, from bottom to top, methods, packages, modules, workflows, coupled workflows and applications. This means, everything is considered a component,

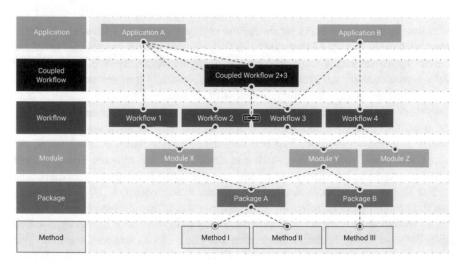

Fig. 5.4 DASF component-based approach on six levels of abstraction

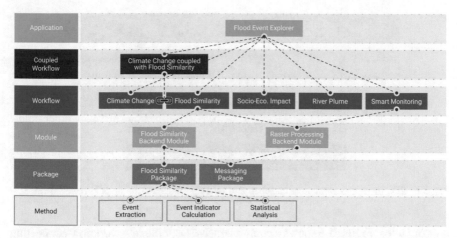

Fig. 5.5 Exemplarily implementation of the general component-based approach: The Flood Similarity Workflow as part of the Flood Event Explorer

while each component uses or accesses various other components from the levels below, indicated by the dashed lines. The usage of individual components is not exclusive, meaning a component can be used by multiple other components from higher levels; for example, Module X and Module Y are using Package A.

An implementation of this general component-based approach is presented in Fig. 5.5. It shows the components involved in the Flood Similarity Workflow as part of the Flood Event Explorer application. It allows the comparison of flood events on the basis of several flood indicators (chapter 5.3.3). On the bottom level, several methods for flood event extraction, event indicator calculation and statistical analysis are implemented, they are combined into packages, and packages are wrapped into modules, which are finally linked by the workflow "Flood Similarity". The Flood Similarity Workflow can also be coupled with other workflows, such as the Climate Change Workflow in the example below. The coupled workflows allow for answering the more complex scientific questions "How could hydro-meteorological controls of flood events develop under projected climate change" (Chapter 5.3.3).

Since all components are independent entities, we need to provide one or several techniques of combining and integrating these components (implementing the dashed lines from Fig. 5.4 and 5.5). For this, we rely on different techniques depending on the abstraction level of the component. The main focus of DASF is the combination of methods to create an integrated workflow. In order to provide a standardized access to any kind of method implementation, we introduce a module layer that wraps packages with their methods. These "wrapper" modules provide a remote procedure call (RPC, White 1976) protocol implementation, harmonizing the access to individual methods across platforms and programming languages. Once the needed methods are provided via individual RPC modules, a workflow can utilize/integrate them by connecting their corresponding inputs and outputs. In the case of other abstraction levels, we rely on programming language-specific integration techniques, like object-oriented

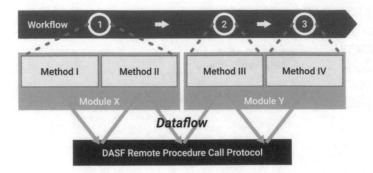

Fig. 5.6 RPC module communication scheme of DASF to integrate methods into workflows

programming. Figure 5.6 shows a comprehensive example of a workflow connecting various methods via RPC modules. Step 1 of the workflow involves the connection of Methods I and II, which are provided by Module X, while Step 2 and 3 are covered by Methods III and IV, respectively. The individual modules expose the methods only through the DASF RPC protocol, so the corresponding input and output data of the combined methods will not be directly passed to one another but always through a connecting mediator.

As discussed above, the DASF RPC approach relies on an appropriate mediator entity passing the actual data between different modules and their exposed methods. In order to support heterogeneous and distributed deployment environments, our RPC communication approach uses a Publish-Subscribe Message Broker technique as a mediator. Within DASF, we use Apache Pulsar [Pulsar 2021] as a ready-to-use message broker implementation. The RPC/Message Broker approach also provides a certain flexibility when it comes to deploy and execute individual components on different IT platforms and systems and to execute an individual component in a distributed IT environment, in our case various Helmholtz centres (compare Fig. 5.3). It also supports reusability and integration of new and existing distributed components and facilitates collaborative development of several methods at the same time, without interfering with each other; the "clear separation of concerns" allows for the design and implementation of any number of components in parallel.

The presented CBSE approach implemented by DASF enables us to address the initially formulated requirements. In contrast to common workflow engines, the shown approach requires some additional effort to implement a workflow and its methods. Yet it provides the necessary flexibility for the highly exploratory data analysis workflows we have in science. In order to show the framework's capabilities, we used it to implement the Digital Earth Flood Event Explorer, presented below. All DASF components (Eggert 2021) are registered in the corresponding language-specific package repositories, like npm (e.g. https://www.npmjs.com/pac kage/dasf-web) and pypi (e.g. https://pypi.org/project/demessaging/). The framework' sources are available via gitlab (https://git.geomar.de/digital-earth/dasf) and licensed under the Apache-2.0 license.

5.3 The Digital Earth Flood Event Explorer—A Showcase for Data Analysis and Exploration with Scientific Workflows

5.3.1 The Showcase Setting

We applied the concept of scientific workflows and the component-based Data Analytics Software Framework (DASF) to an exemplary showcase, the Digital Earth Flood Event Explorer (Eggert et al. 2022). The Flood Event Explorer should support geoscientists and experts to analyse flood events along the process cascade event generation, evolution and impact across atmospheric, terrestrial and marine disciplines. It aims at answering the following geoscientific questions:

- How does precipitation change over the course of the twenty-first century under different climate scenarios over a certain region?
- What are the main hydro-meteorological controls of a specific flood event?
- What are useful indicators to assess socio-economic flood impacts?
- How do flood events impact the marine environment?
- What are the best monitoring sites for upcoming flood events?

Our aim was to develop scientific workflows providing enhanced analysis methods from statistics, machine learning and visual data exploration that are implemented in different languages and software environments, and that access data from a variety of distributed databases. Within the showcase, we wanted to investigate how the concept of scientific workflows and component-based software engineering can be adapted to a "real-world" setting and what the benefits and limitations are.

We chose the Elbe River in Germany as a concrete test site since data are available for several severe and less severe flood events in this catchment. The collaborating scientists are from different Helmholtz research centres and belong to different scientific fields such as hydrology, climate, marine, and environmental science, and computer science and data science.

5.3.2 Developing and Implementing Scientific Workflows for the Flood Event Explorer

We developed and implemented scientific workflows for each question that has to be answered within the Flood Event Explorer; Figure 5.7 gives an overview about the workflows.

We established interdisciplinary teams with geo- and computer scientists to provide all the particular expertise that is needed. The development and implementation of the scientific workflows was an iterative co-design process. Geoscientists developed the conceptual workflows. They had to deal with the following

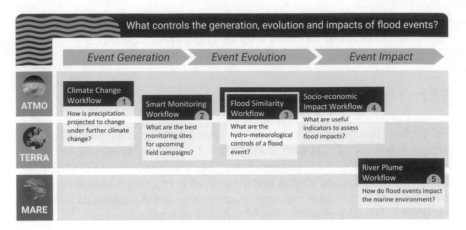

Fig. 5.7 Overview on the workflows of the Digital Earth Flood Event Explorer

issues: What analytical approaches are suitable to answer the geoscientific questions? What data and analysis tasks and methods are needed? What enhanced methods can improve traditional analysis approaches? The conceptual workflows are documented in flowcharts; chapter 5.3.3 gives an overview about it. Computer scientists had to contribute their particular expertise: What approaches are suitable to integrate the analysis methods? How to realize the highly interactive exploration of data and results? How to enable reusability of the methods? They developed the Data Analytics Software Framework (DASF) with its integration and visualization module (chapter 5.2.3 and 3.2) to implement exploratory scientific workflows. They also transformed the conceptual workflows into digital ones.

Additionally, geo- and data scientists developed and implemented several enhanced analysis methods as part of the workflows that go beyond traditional analysis approaches. They are described in this book in chapters 3.2 and 4.5.1.

A further idea we followed in the showcase was to combine single workflows to answer more complex geoscientific questions within the Flood Event Explorer. By coupling data and methods of single analysis workflows into a larger data analysis chain—a combined workflow—we want to integrate knowledge from different scientific communities to give answers to more complex questions that go beyond single perspectives. Combined workflows have the potential to facilitate a more comprehensive view to processes in the Earth system, in our case flood events. Based on the conceptual workflow descriptions, we identified the following three complex questions that can be answered by combining single workflows:

- How could hydro-meteorological controls of flood events develop under projected climate change?
- Can large-scale flood events be detected as exceptional nutrient inputs that lead to algae blooms in the North Sea?
- How should an optimal future groundwater monitoring network be designed by incorporating future climate scenarios?

The flowcharts presented at the end of chapter 5.3.3 show the combination of single workflows to give answers to the questions.

5.3.3 The Workflows of the Flood Event Explorer

This chapter gives a brief overview about the workflows we have developed for the Flood Event Explorer. The first part presents the workflows shown in Fig. 5.7 with (a) a short textual description of each workflow's objective, functionality and added value, (b) the flowcharts of the conceptual workflows and (c) the visual user interface of the digital workflows. The second part introduces the combined workflows that allow us to answer more complex geoscientific questions.

Access to all workflows as well as their documentation, additional media and references is provided via a mutual landing page (http://rz-vm154.gfz-potsdam.de:8080/de-flood-event-explorer/). The landing page provides an interactive overview of the environmental compartments along the process chain of flood events and the involved workflows.

5.3.3.1 The Climate Change Workflow

The goal of the Climate Change Workflow (Fig. 5.8, Fig. 5.9) is to support the analysis of climate-driven changes in flood-generating climate variables, such as precipitation or soil moisture, using regional climate model simulations from the Earth System Grid Federation (ESGF) data archive. It should support to answer the geoscientific question "How does precipitation change over the course of the 21st century under different climate scenarios, compared to a 30-year reference period over a certain region?"

Extraction of locally relevant data over a region of interest (ROI) requires climate expert knowledge and data processing training to correctly process large ensembles of climate model simulations; the Climate Change Workflow tackles this problem. It supports scientists to define the regions of interest, customize their ensembles from the climate model simulations available on the Earth System Grid Federation (ESGF) and define variables of interest and relevant time ranges.

The Climate Change Workflow provides: (1) a weighted mask of the ROI; (2) weighted climate data of the ROI; (3) time series evolution of the climate over the ROI for each ensemble member; (4) ensemble statistics of the projected change; and lastly, (5) an interactive visualization of the region's precipitation change projected by the ensemble of selected climate model simulations for different Representative Concentration Pathways (RCPs). The visualization includes the temporal evolution of precipitation change over the course of the twenty-first century and statistical characteristics of the ensembles for two selected 30-year time periods for the mid- and the end of the twenty-first century (e.g. median and various percentiles).

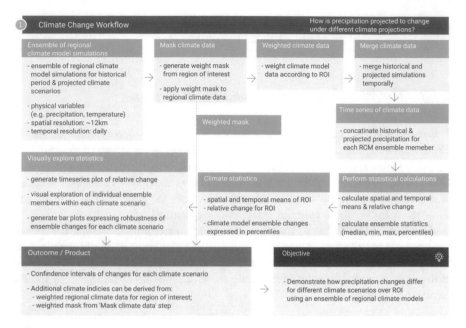

Fig. 5.8 Conceptual workflow description of the Climate Change Workflow

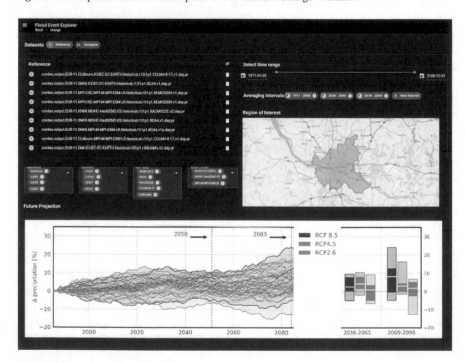

Fig. 5.9 Visual User Interface of the Climate Change Workflow

The added value of the Climate Change Workflow is threefold. First, there is a reduction in the number of different software programs necessary to extract locally relevant data. Second, the intuitive generation and access to the weighted mask allow for the further development of locally relevant climate indices. Third, by allowing access to the locally relevant data at different stages of the data processing chain, scientists can work with a vastly reduced data volume allowing for a greater number of climate model ensembles to be studied, which translates into greater scientific robustness. Thus, the Climate Change Workflow provides much easier access to an ensemble of high-resolution simulations of precipitationover a given ROI, presenting the region's projected precipitation change using standardized approaches, and supporting the development of additional locally relevant climate indices.

Additional Climate Change Workflow Media

Workflow-In-Action Video:
    https://youtu.be/MuX-oH1W2wk
    Conceptual workflow description Chart:
    http://rz-vm154.gfz-potsdam.de:8080/de-flood-event-explorer/images/1_clim
ate_change_flowchart.svg
    Source Code Repository:
    https://git.geomar.de/digital-earth/flood-event-explorer/fee-climate-change-wor
kflow
    Registered Software DOIs:
    de-esgf-download (https://doi.org/10.5281/zenodo.5793278)
    dc-climate-change-analysis (https://doi.org/10.5281/zenodo.5833043)
    Digital Earth Climate Change Backend Module (https://doi.org/10.5281/zenodo.
5833258)
    The Climate Change Workflow (https://doi.org/10.5880/GFZ.1.4.2022.003)
    Accessible via the Flood Event Explorer Landing Page:
    http://rz-vm154.gfz-potsdam.de:8080/de-flood-event-explorer/

5.3.3.2 The Flood Similarity Workflow

River floods and associated adverse consequences are caused by complex interactions of hydro-meteorological and socio-economic preconditions and event characteristics. The Flood Similarity Workflow (Fig. 5.10, Fig. 5.11) supports the identification, assessment and comparison of hydro-meteorological controls of flood events.

The analysis of flood events requires the exploration of discharge time series data for hundreds of gauging stations and their auxiliary data. Data availability and accessibility and standard processing techniques are common challenges in that application and addressed by this workflow.

The Flood Similarity Workflow allows the assessment and comparison of arbitrary flood events. The workflow includes around 500 gauging stations in Germany

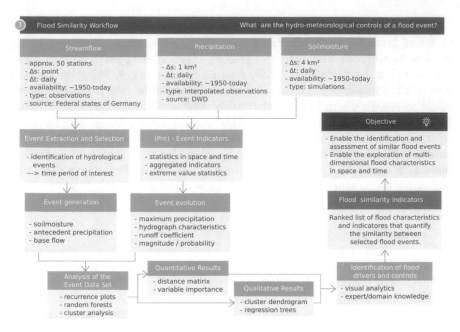

Fig. 5.10 Conceptual workflow description of the Flood Similarity Workflow

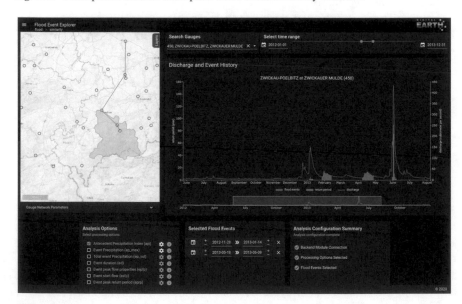

Fig. 5.11 (Parts of the) Visual User Interface of the Flood Similarity Workflow

comprising discharge data and the associated extreme value statistics as well as precipitation and soil moisture data. This provides the basis to identify and compare flood events based on antecedent catchment conditions, catchment precipitation, discharge hydrographs and inundation maps. The workflow also enables the analysis of multidimensional flood characteristics including aggregated indicators (in space and time), spatial patterns and time series signatures.

The added value of the Flood Event Explorer comprises two major points. First, scientists work on a common, homogenized database of flood events and their hydro-meteorological controls for a large spatial and temporal domain, with fast and standardized interfaces to access the data. Second, the standardized computation of common flood indicators allows a consistent comparison and exploration of flood events.

Additional Flood Similarity Workflow Media

Workflow-In-Action Video:
   https://youtu.be/3Z-3oyu8bP4
   Conceptual workflow description Chart:
   http://rz-vm154.gfz-potsdam.de:8080/de-flood-event-explorer/images/2_f
lood_similarity_flowchart.svg
   Source Code Repository:
   https://git.geomar.de/digital-earth/flood-event-explorer/fee-flood-similarity-wor
kflow
   Registered Software DOIs:
   Digital Earth Similarity Backend Module (https://doi.org/10.5281/zenodo.580
1319)
   The Flood Similarity Workflow (https://doi.org/10.5880/GFZ.1.4.2022.003)
   Accessible via the Flood Event Explorer Landing Page:
   http://rz-vm154.gfz-potsdam.de:8080/de-flood-event-explorer/

5.3.3.3 The Socio-Economic Flood Impacts Workflow

The Socio-Economic Flood Impacts Workflow (Fig. 5.12) aims to support the identification of relevant controls and useful indicators for the assessment of flood impacts. It should support answering the question "What are useful indicators to assess socio-economic flood impacts?". Floods impact individuals and communities and may have significant social, economic and environmental consequences. These impacts result from the interplay of hazard—the meteo-hydrological processes leading to high water levels and inundation of usually dry land; exposure—the elements affected by flooding such as people, build environment or infrastructure; and vulnerability—the susceptibility of exposed elements to be harmed by flooding.

In view of the complex interactions of hazard and impact processes, a broad range of data from disparate sources need to be compiled and analysed across the

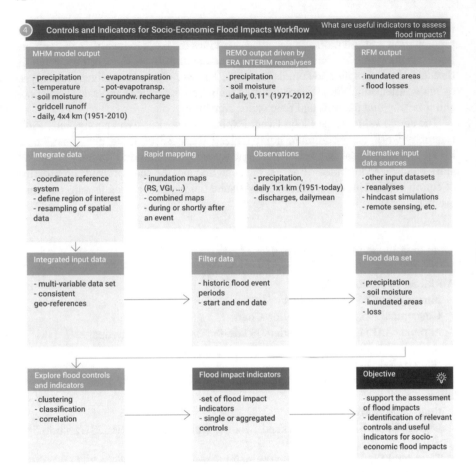

Fig. 5.12 Conceptual workflow description of the Socio-Economic Flood Impacts Workflow

boundaries of climate and atmosphere, catchment and river network, and socio-economic domains. The workflow approaches this problem and supports scientists to integrate observations, model outputs and other datasets for further analysis in the region of interest.

The workflow provides functionalities to select the region of interest, access hazard, exposure and vulnerability-related data from different sources, identifying flood periods as relevant time ranges, and calculate defined indices. The integrated input dataset is further filtered for the relevant flood event periods in the region of interest to obtain a new comprehensive flood dataset. This spatio-temporal dataset is analysed using data science methods such as clustering, classification or correlation algorithms to explore and identify useful indicators for flood impacts. For instance, the importance of different factors or the interrelationships amongst multiple variables to shape flood impacts can be explored.

The added value of the Socio-Economic Flood Impacts Workflow is twofold. First, it integrates scattered data from disparate sources and makes it accessible for further analysis. As such, the effort to compile, harmonize and combine a broad range of spatio-temporal data is clearly reduced. Also, the integration of new datasets from additional sources is much more straightforward. Second, it enables a flexible analysis of multivariate data, and by reusing algorithms from other workflows, it fosters a more efficient scientific work that can focus on data analysis instead of tedious data wrangling.

Additional Socio-Economic Flood Impacts Workflow Media

Conceptual workflow description Chart:
   http://rz-vm154.gfz-potsdam.de:8080/de-flood-event-explorer/images/4_i
mpact_flowchart.svg
   Source Code Repository:
   https://git.geomar.de/digital-earth/flood-event-explorer/fee-socio-impact-wor
kflow
   Registered Software DOIs:
   Digital Earth 'Controls and Indicators for Flood Impacts' Backend Module
(https://doi.org/10.5281/zenodo.5801815)
   The Socio-Economic Flood Impacts Workflow (https://doi.org/10.5880/GFZ.1.4.
2022.005)
   Accessible via the Flood Event Explorer Landing Page:
   http://rz-vm154.gfz-potsdam.de:8080/de-flood-event explorer/

5.3.3.4 The River Plume Workflow

The focus of the River Plume Workflow (Fig. 5.13, Fig. 5.14) is the impact of riverine flood events on the marine environment. At the end of a flood event chain, an unusual amount of nutrients and pollutants is washed into the North Sea, which can have consequences, such as increased algae blooms. The workflow aims to enable users to detect a river plume in the North Sea and to determine its spatio-temporal extent.
   Identifying river plume candidates can either happen manually in the visual inter-face (chapter 3.2) or also through an automatic anomaly detection algorithm, using Gaussian regression (chapter 4.5.1). In both cases, a combination of observational data, namely FerryBox transects and satellite data, and model data is used. Once a river plume candidate is found, a statistical analysis supplies additional detail on the anomaly and helps to compare the suspected river plume to the surrounding data. Simulated trajectories of particles starting on the FerryBox transect at the time of the original observation and modelled backwards and forwards in time help to verify the origin of the river plume and allow users to follow the anomaly across the North Sea. An interactive map enables users to load additional observational data into the workflow, such as ocean colour satellite maps, and provides them with an overview

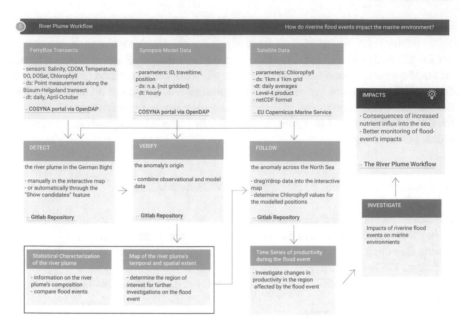

Fig. 5.13 Conceptual workflow description of the River Plume Workflow

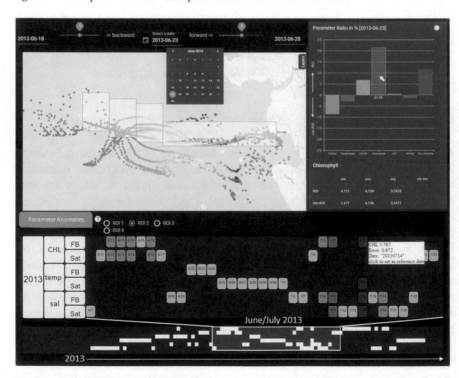

Fig. 5.14 Visual User Interface of the River Plume Workflow

of the flood impacts and the river plume's development on its way through the North Sea. In addition, the workflow offers the functionality to assemble satellite-based chlorophyll observations along model trajectories as time series. They allow scientists to understand processes inside the river plume and to determine the timescales on which these developments happen. For example, chlorophyll degradation rates in the Elbe River plume are currently investigated using these time series.

The workflow's added value lies in the ease with which users can combine observational FerryBox data with relevant model data and other datasets of their choice. Furthermore, the workflow allows users to visually explore the combined data and contains methods to find and highlight anomalies. The workflow's functionalities also enable users to map the spatio-temporal extent of the river plume and investigate the changes in productivity that occur in the plume. All in all, the River Plume Workflow simplifies the investigation and monitoring of flood events and their impacts in marine environments.

Additional River Plume Workflow Media

Workflow-In-Action Video:
   https://youtu.be/yl8ngubBxYY
   Conceptual workflow description Chart:
   http://rz-vm154.gfz-potsdam.de:8080/de-flood-event-explorer/images/5_river_plume_flowchart.svg
   Source Code Repository:
   https://git.geomar.de/digital-earth/flood-event-explorer/fee-river-plume-workflow
   Registered Software DOIs:
   The River Plume Workflow (https://doi.org/10.5880/GFZ.1.4.2022.006)
   Accessible via the Flood Event Explorer Landing Page:
   http://rz-vm154.gfz-potsdam.de:8080/de-flood-event-explorer/

5.3.3.5 The Smart Monitoring Workflow

A deeper understanding of the Earth system as a whole and its interacting sub-systems depends not only on accurate mathematical approximations of the physical processes but also on the availability of environmental data across time and spatial scales. Even though advanced numerical simulations and satellite-based remote sensing in conjunction with sophisticated algorithms such as machine learning tools can provide 4D environmental datasets, local and mesoscale measurements continue to be the backbone in many disciplines such as hydrology. Considering the limitations of human and technical resources, monitoring strategies for these types of measurements should be well designed to increase the information gain provided. One helpful set of tools to address these tasks is data exploration frameworks providing qualified data from different sources and tailoring available computational and visual methods to explore and analyse multi-parameter datasets. In this context, we developed a Smart Monitoring Workflow (Fig. 5.15, Fig. 5.16) to determine the most suitable time and location for event-driven, ad hoc monitoring in hydrology using soil moisture measurements as our target variable.

The Smart Monitoring Workflow (Nixdorf et al. 2022) consists of three main steps. First is the identification of the region of interest, either via user selection or recommendation based on spatial environmental parameters provided by the user. Statistical filters and different colour schemes can be applied to highlight different

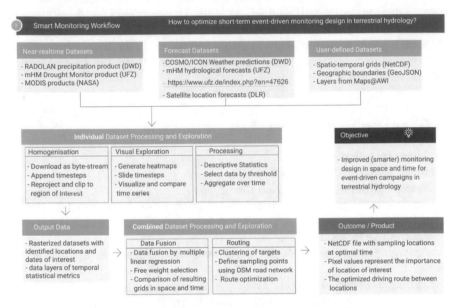

Fig. 5.15 Conceptual workflow description of the Smart Monitoring Workflow

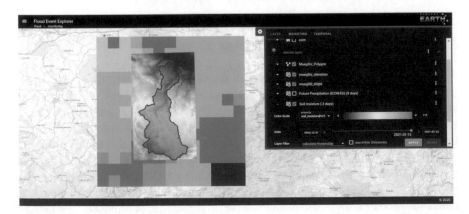

Fig. 5.16 Visual User Interface of the Smart Monitoring Workflow (from Nixdorf et al, 2022)

regions. The second step is accessing time-dependent environmental parameters (e.g. rainfall and soil moisture estimates of the recent past, weather predictions from numerical weather models and swath forecasts from Earth observation satellites) for the region of interest and visualizing the results. Lastly, a detailed assessment of the region of interest is conducted by applying filter and weight functions in combination with multiple linear regressions on selected input parameters. Depending on the measurement objective (e.g. highest/lowest values, highest/lowest change), most suitable areas for monitoring will subsequently be visually highlighted. In combination with the provided background map, an efficient route for monitoring can be planned directly in the exploration environment.

The added value of the Smart Monitoring Workflow is multifold. The workflow gives the user a set of tools to visualize and process their data on a background map and in combination with data from public environmental datasets. For raster data from public databases, tailor-made routines are provided to access the data in the spatio-temporal limits required by the user. Aiming to facilitate the design of terrestrial monitoring campaigns, the platform and device-independent approach of the workflow give the user the flexibility to design a campaign at the desktop computer first and to refine it later in the field using mobile devices. In this context, the ability of the workflow to plot time series of forecast data for the region of interest empowers the user to react quickly to changing conditions, e.g. thunderstorm showers, by adapting the monitoring strategy, if necessary.

Finally, the integrated routing algorithm assists to calculate the duration of a planned campaign as well as the optimal driving route between often scattered monitoring locations.

Additional Smart Monitoring Workflow Media

Workflow-In-Action Video:
    https://youtu.be/m5ivu86kpfg
    Conceptual workflow description Chart:
    http://rz-vm154.gfz-potsdam.de:8080/de-flood-event-explorer/images/3_s mart_monitoring_flowchart.svg
    Source Code Repository:
    https://git.geomar.de/digital-earth/flood-event-explorer/fee-smart-monitoring-workflow
    Registered Software DOIs:
    Digital Earth Smart Monitoring Backend Module (Tocap) (https://doi.org/10.5281/zenodo.5824566)
    The Smart Monitoring Workflow (Tocap) (https://doi.org/10.5880/GFZ.1.4.2022.004)
    Accessible via the Flood Event Explorer Landing Page:
    http://rz-vm154.gfz-potsdam.de:8080/de-flood-event-explorer/

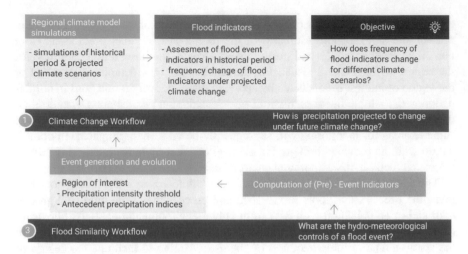

Fig. 5.17 Conceptual workflow description of the combined Flood Similarity and Climate Change Workflow

5.3.3.6 Workflow Combinations

The combination of scientific workflows to foster data analysis and knowledge creation across disciplines was one idea we followed in Digital Earth. The following examples show how the workflows presented above can be combined to give answers to more complex questions.

Combination of the Flood Similarity Workflow and the Climate Change Workflow

A combination of the Flood Similarity Workflow and the Climate Change Workflow (Fig. 5.17) can give an answer to the question "How could hydro-meteorological controls of flood events develop under projected climate change?". The Flood Similarity Workflow provides a set of flood indices that allow for a comparison of flood events. These indices describe the antecedent catchment soil moisture and catchment precipitation, amongst other properties. The Flood Similarity Workflow determines the flood indices for the historical and present period for a region of interest. These flood indices are then calculated from climate projections from the Climate Change Workflow. The Climate Change Workflow provides information for projected future conditions regarding how the precipitation and soil moisture characteristics change under different future scenarios for mid- and end of the twenty-first century over a region of interest. The combined workflows show how controls of flood events, described by the indices, develop and how the probability of floods of a given magnitude changes over a certain region.

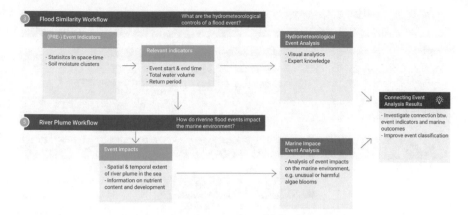

Fig. 5.18 Conceptual workflow description of the combined River Plume and Flood Similarity Workflow

Combination of the Flood Similarity Workflow and the River Plume Workflow

Combining the Flood Similarity Workflow and the River Plume Workflow (Fig. 5.18) addresses the question, whether large-scale flood events can be detected as exceptional nutrient inputs into the North Sea that lead to algae blooms. Next to the start and end dates of the flood event, the Flood Similarity Workflow provides characteristics like the return period and the total water volume during the event. Given the temporal and hydrological characteristics, the River Plume Workflow evaluates the impact of fluvial floods on water quality anomalies in the North Sea. The combination of both workflows allows investigating the connection between flood event indicators and the marine impacts caused by said flood event. Results from this analysis can be used to improve the classification of flood events in future.

Combination of the Climate Change and the Smart Monitoring Workflow

Groundwater monitoring stations are essential for protecting the groundwater from harmful pollutants and ensuring good water quality. As precipitation and temperature patterns change under different climate evolutions, so will groundwater recharging rates and quality. These changes need to be monitored adequately to ensure that countermeasures start on time and are efficient. Currently, groundwater monitoring wells are unevenly distributed across the German aquifer systems. By combining the Climate Change Workflow and the Smart Monitoring Workflow (Fig. 5.19), we can assist to improve the groundwater monitoring network design required in future for different climate projections.

The Climate Change Workflow provides information on precipitation changes over a region of interest based on regional climate model projections at a resolution

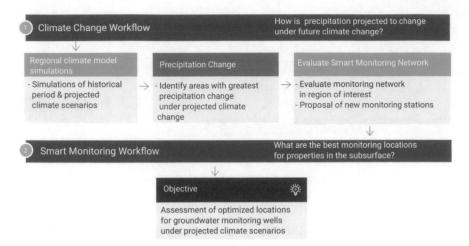

Fig. 5.19 Conceptual workflow description of the combined Climate Change and Smart Monitoring Workflow

of 144 km². These data help to assess where the largest climate-driven changes in groundwater resources are expected and combined with the Smart Monitoring Workflow; it can determine whether there are a sufficient number of groundwater monitoring wells installed and can further assist in determining suitable locations for the installation of new monitoring wells.

5.4   Assessment of the Scientific Workflow Concept

Our assessment of the concept of scientific workflows focuses on the challenges and needs we addressed in chapter 5.1. Along these challenges, we present our experiences with benefits, limitations and efforts of scientific workflows.

Challenge 1: Seamless data analysis chains to extensively analyse and explore spatio-temporal and multivariate data.

As we could show in our Flood Event Explorer, scientific workflows provide an analysis environment that integrates different analysis methods and tools into seamless data analysis chains and facilitates integration of data from various sources; the visual front ends of the digital workflows enable detailed exploration of the multivariate spatio-temporal data. This is a clear benefit. In more detail, our exemplary workflows of the Flood Event Explorer attain the following benefits:

- Reduction in the number of different software programs;
- Much easier access to data and ensembles of simulations;
- Ease with which users can combine datasets of their choice;
- Intuitive generation and access to spatial or temporal subsets of data allow for further analysis;

- By allowing access to spatial subsets of data (e.g. locally relevant data) at different stages of the data processing chain, scientists can work with a vastly reduced data volume allowing for a more comprehensive analysis (e.g. a greater number of model ensembles to be studied), which translates into greater scientific robustness.

Scientific workflows have proven to be helpful means to create improved work environments with seamless data analysis chains to extensively analyse and explore spatio-temporal and multivariate data.

Besides this benefit, scientific workflows have one major limitation; this is the effort for its development and implementation. The use of workflow engines and the manual implementation of workflows need additional effort and expertise. The manual implementation we have chosen in the Digital Earth project requires knowledge from software engineering and consequently various expertise. Although our Data Analytics Software Framework (DASF) allows reuse of workflow components and thus eases the implementation process, a minimum of software development expertise is still required.

Another limitation of our scientific workflows is their predefined data analysis approach with a set of predefined analysis methods. Due to the component-based approach of our Data Analytics Software Framework (DASF), methods can be substituted easily; however, this also needs particular expertise in software development.

Challenge 2: Sharing and reuse of methods and software tools.

The systematic description of scientific workflows and its implementation with the Data Analytics Software Framework (DASF) enable sharing and reuse of all workflows and workflow components we have developed for the Flood Event Explorer. Several of our workflow components (data analysis and visualization methods) could already be used again in other projects without any difficulties and so could save time and costs for reimplementation. Examples are the GFZ Earthquake Explorer (https://geofon.gfz-potsdam.de/eqexplorer/test/) or the Geochemical Explorer (http://rz-vm154.gfz-potsdam.de:8080/gcex/). The DASF also allowed reusing already existing data analysis software and integrating it into the workflows of the Flood Event Explorer independent of the programming language. A further benefit is the sharing of extensive analysis approaches with other scientists. Scientific workflows go beyond sharing single methods and tools; they implement whole analysis approaches and so support more standardized and comparable analysis and results. Examples within our Flood Event Explorer are the approaches to determine projected precipitation change of defined regions, to generate common flood indicators for a consistent comparison and exploration of flood events or to identify the best monitoring sites and time (chapter 5.3.3). Conceptual workflow descriptions can also be used as metadata providing suitable information for reuse and sharing.

The benefits that scientific workflows and the Data Analytics Software Framework (DASF) provide for sharing and reuse of single analysis methods and extensive analysis approaches require some additional effort and work. First the RPC wrapper module, described in chapter 5.2.3, has to be added to the actual method. The effort to do so depends on the used programming language. In case of the widely used Python programming language, we provide an easy-to-use annotation-based interface for this. Furthermore, a message broker instance, as described in chapter 5.2.3, has to be provided and maintained. Finally, the module wrapping the developed method has to be deployed on a suitable host. In addition to the technical effort, a shared, reusable module needs to be documented properly to ease the workload of the reusing party. In general, we consider reuse and sharing of methods on the module level. However, this always requires the use of the DASF in the scope of the reusing context. Since we applied CBSE on all levels of abstraction, the individual methods could also directly be reused without the advantages and disadvantages of DASF.

Challenge 3: Communicate and combine scientific approaches across disciplines.

The conceptual workflows we described in flow charts are a systematic documentation of scientific knowledge and approaches; they show how scientists proceed to answer a geoscientific question. In the Digital Earth project, it became very clear that conceptual workflow descriptions can serve as significant means to understand approaches of other scientists and to discuss how various approaches can be combined. Examples in our Flood Event Explorer are the combined workflows presented in chapter 5.3.3. Once it is clear on the conceptual level how scientific approaches and workflows can be combined, they also can be linked on the implementation level for the digital workflows. The concept of scientific workflows supports linking approaches across disciplines on the conceptual as well as on the digital level.

Our experience also showed that describing conceptual scientific workflows is not an easy task for geoscientists. The systematic documentation of scientific approaches needs some experience. Scientists have to learn to represent their approaches in a systematic way either in flow charts or in other forms; this requires additional effort from geoscientists.

Challenge 4: Suitable interface between geo- and computer science.

The conceptual workflow descriptions largely support the communication between geo- and computer scientists; they are recognizable means of translation between the two worlds. The geo- and computer scientists collaborating in Digital Earth assessed the task analysis and flowcharts to elicitate and document the conceptual workflows as essential for collaboration. Geoscientists are forced to systematically capture their requirements and analytical tasks; computer scientists receive a sound basis to map analytical tasks to suitable methods that fulfil the task in the digital workflows.

To summarize: The experience we made in Digital Earth with adopting the concept of scientific workflows to geosciences can be summarized as following: The concept of scientific workflows provides a suitable frame to tackle the needs we have addressed for data analysis and exploration in Digital Earth: digital workflows enable seamless executable data analysis chains (challenge 1), and they also support sharing and reuse of data analysis methods and tools (challenge 2). Conceptual workflow descriptions are suitable means to exchange and combine scientific approaches and knowledge (challenge 3); they also serve as a suitable interface between approaches from geo- and computer science (challenge 4). To utilize this potential of workflows, some additional effort is necessary. First, geoscientists have to force thinking in workflows and gain skills and experience in describing conceptual scientific workflows. Second, state-of-the-art concepts from computer science are needed to develop and implement highly explorative digital scientific workflows. Third, a close collaboration and co-design of geo- and computer scientists is required to develop suitable scientific workflows. This requires people who are willing and able for border crossing and thinking out of the box. To shape such people was one aim and success of the Digital Earth project.

References

Atkinson M, Geins S, Montagnat J, Taylor I (2017) Scientific workflows: past, present, future. Futur Gener Comput Syst 75(2017):216–227

Cerezo N, Montagnat J, Blay-Fornarino M (2013) Computer-assisted scientific workflow design. J Grid Computing 11:585–612. https://doi.org/10.1007/s10723-013-9264-5

Dijkstra, Edsger W (1982) On the role of scientific thought. Selected writings on computing: a personal perspective. New York, NY, USA: Springer-Verlag. pp 60–66. ISBN 0-387-90652-5

Eggert D, Dransch D (2021) DASF: a data analytics software framework for distributed environments. GFZ Data Services. https://doi.org/10.5880/GFZ.1.4.2021.004

Eggert D, Rabe D, Dransch D, Lüdtke S, Schröter K, Nam C, Nixdorf E, Wichert V, Abraham N, Merz B (2022) Digital Earth Flood Event Explorer: a showcase for data analysis and exploration with scientific workflows. GFZ Data Services. https://doi.org/10.5880/GFZ.1.4.2022.001

Heineman, George T and Councill, William T (2001) Component-based software engineering: putting the pieces together. Addison-Wesley Professional, Reading 2001 ISBN 0-201-70485-4

Hollingsworth D (1995). Workflow Management Coalition: the Workflow Reference Model.

Jonassen DH, Tessmer M, Hannum WH (eds) (1998) Task analysis methods for instructional design, Routledge, Taylor and Francis Group

McIlroy, Malcolm Douglas (1969, January). Mass produced software components (PDF). Software engineering: report of a conference sponsored by the NATO Science Committee, Garmisch, Germany, 7–11 October 1968. Scientific Affairs Division, NATO, p 79

Nixdorf E, Eggert D, Morstein P, Kalbacher T, Dransch D (2022) Tocap: a web tool for ad-hoc campaign planning in terrestrial hydrology. J Hydroinformatic. https://doi.org/10.2166/hydro.2022.057

Pulsar2021: https://pulsar.apache.org/

Schraagen JM, Chipman SF, Valerie L, Shalin VL (eds) (2000) Cognitive task analysis, Psychology Press

White, James E (1976, January 14) RFC 707. A high-level framework for network-based resource sharing. Proceedings of the 1976 National Computer Conference

Workflow Systems (2021) https://galaxyproject.org/, https://kepler-project.org/, https://taverna.incubator.apache.org/, https://pegasus.isi.edu/

Chapter 6
The Digital Earth Smart Monitoring Concept and Tools

Uta Koedel, Peter Dietrich, Philipp Fischer, Jens Greinert, Ulrich Bundke,
Ewa Burwicz-Galerne, Antonie Haas, Isabel Herrarte, Amir Haroon,
Marion Jegen, Thomas Kalbacher, Marcel Kennert, Tobias Korf,
Ralf Kunkel, Ching Yin Kwok, Christoph Mahnke, Erik Nixdorf,
Hendrik Paasche, Everardo González Ávalos, Andreas Petzold,
Susanne Rohs, Robert Wagner, and Andreas Walter

Abstract Reliable data are the base of all scientific analyses, interpretations and
conclusions. Evaluating data in a smart way speeds up the process of interpretation
and conclusion and highlights where, when and how additionally acquired data in
the field will support knowledge gain. An extended SMART monitoring concept is
introduced which includes SMART sensors, DataFlows, MetaData and Sampling
approaches and tools. In the course of the Digital Earth project, the meaning of
SMART monitoring has significantly evolved. It stands for a combination of hard-
and software tools enhancing the traditional monitoring approach where a SMART
monitoring DataFlow is processed and analyzed sequentially on the way from the
sensor to a repository into an integrated analysis approach. The measured values
itself, its metadata, and the status of the sensor, and additional auxiliary data can be
made available in real time and analyzed to enhance the sensor output concerning

U. Koedel (✉) · P. Dietrich · T. Kalbacher · C. Y. Kwok · E. Nixdorf · H. Paasche
Helmholtz Centre for Environmental Research—UFZ, Leipzig, Germany
e-mail: uta.koedel@ufz.de

P. Fischer · A. Haas · I. Herrarte · A. Walter
Alfred Wegener Institute Helmholtz Centre for Polar and Marine Research, Bremerhaven,
Germany

J. Greinert · A. Haroon · M. Jegen · E. González Ávalos
GEOMAR Helmholtz Centre for Ocean Research Kiel, Kiel, Germany

U. Bundke · M. Kennert · T. Korf · R. Kunkel · C. Mahnke · A. Petzold · S. Rohs
Forschungszentrum Jülich GmbH, Jülich, Germany

R. Wagner
Leibniz Institute for Baltic Sea Research Warnemünde IOW, Rostock, Germany

E. Nixdorf
Department of Groundwater and Soil, Federal Institute for Geosciences and Natural Resources
(BGR), Hannover, Germany

E. Burwicz-Galerne
University of Bremen, Bremen, Germany

© The Author(s) 2022   85
L. M. Bouwer et al. (eds.), *Integrating Data Science and Earth Science*,
SpringerBriefs in Earth System Sciences,
https://doi.org/10.1007/978-3-030-99546-1_6

accuracy and precision. Although several parts of the four tools are known, technically feasible and sometimes applied in Earth science studies, there is a large discrepancy between knowledge and our derived ambitions and what is feasible and commonly done in the reality and in the field.

Keywords Adaptive · Prediction · Monitoring · Sensors · Metadata · FAIR · DataFlow · SMART concept · SMART tools

6.1 Challenges

The understanding of the Earth system with all its different habitats, processes, connections between spheres and feedback loops demands the repeated observation of high number of parameters in high spatial and temporal resolution over long time. Such kind of monitoring is e.g. the base of our understanding of climate change, changes in biodiversity or on shorter time scales e.g. the development of a sediment plume in the deep sea during deep sea mining. The base of all this knowledge gain is monitoring of specific parameters in the field, be it in the atmosphere, the oceans or on land. Conservation and long-term protection of the environment requires a better understanding of the ecosystem through cross-domain integration of data and knowledge from different disciplines. Current methods used in applied environmental research and scientific surveys are often not sufficient to appropriately address the heterogeneity and dynamic of ecosystem changes.

Thus, new technologies and methods for integrated in-situ and near real-time monitoring with increased spatiotemporal resolution and adaptive functionalities are needed. Recent developments in digital information processing, the internet of things (IoT) or the improved analysis of complex datasets are opening up new possibilities for data-based environmental research. Moreover, these rapidly developing fields call for a paradigm shift towards a SMART monitoring concept that even stronger couples modelling and data acquisition in the field. Having the none achievable goal of "measuring everything, everywhere at any time" in mind sets some challenges to the task of twenty-first-century environmental monitoring which are:

- SMART sensors: Advancing and developing sensors that have real-time data (pre)processing capacities and are linked in a self-organizing sensor network is still a challenging technological task. Automated event-detection, drift correction and failure detection are possible but still rarely done. Real-time data connections and centralized visualization and analyses are more and more established, but the real challenge is that such SMART sensors and sensor networks become easy to use and the standard way of acquiring multiparameter data in the field.
- SMART DataFlow: An easy to use, scalable and adaptable way of receiving data from sensors and re-distributing them through various channels and means also in real time is the challenge for an efficient SMART monitoring DataFlow. Standardized and largely automated procedures are needed to obtain reliable data.

As an essential part of the live cycle of data is the DataFlow crucial for acquiring high quality data at the right time and location

- SMART MetaData: Columns of numbers of a time series alone are not useful without the context these numbers have been generated. The suitable description of data is a prerequisite for any secondary use of data. Apart from FAIR descriptions, the data trustworthiness also needs to be assessed and described to allow a correct evaluation of the data. Compiling these data in a complete manner and raise the awareness again, that MetaData are crucial for the correct use of data, is the real challenge for SMART MetaData.

- SMART Sampling: Objectively finding the best possible sample location in space and time (most informative information for the respective research question), ideally in an automated and adapting way is a challenging task. SMART sampling strategies are supporting this challenge. Applying state-of-the-art statistic and AI methods jointly with interactive visualization and analyses is increasing in the community. The challenge is to spread the knowledge about these methods and present easy ways of using them to lower the hurdle of their application.

Addressing these challenges was the main objective of the SMART monitoring efforts within the Digital Earth project. The involved research centres started, iterated and further developed the idea of a SMART monitoring concept, that finally integrates four conceptual groups of tools, each tackling one of the above stated challenges.

6.2 SMART Monitoring Concept

6.2.1 An Expanded SMART Monitoring Concept

SMART monitoring typically refers to "Self-Monitoring, Analysis, and Reporting Technology" which implies that sensors utilizes e.g. artificial intelligence and big data analysis capabilities to provide an automated data acquisition, simultaneous processing, standardized storage and retrieval of multiple data (Ullo & Sinha 2020; Zhang et al. 2015; Spencer et al. 2004; Thakur et al. 2019; Alharbi & Soh 2019, Lombard et al., 2019). We would like to *expand* the meaning of SMART monitoring in such a way that measured environmental parameters and their values need to be:

- **Specific/Scalable**—Specific relates to accurate and precise and means also that something is clearly defined or identified. Scalable refers to a hierarchical monitoring approach combining multiple sensors measuring across scales and parameters.

- **Measurable/Modular**—Measurable values imply that quantitative information can be measured. Modular refers to a portfolio of different independent methods available to measure a specific parameter to eliminate specific methods' disadvantages/shortcomings.

88 U. Koedel et al.

- **Accepted/Adaptive**—Accepted values are internationally defined in UNESCO standards for the target parameters. Adaptive refers to an easy application to new research questions and the combination of various sensors in a network.
- **Relevant/Robust**—Relevant values are commonly accepted as representative of a specific measurement. Robust refers to self-repair calibration mechanisms in operation in case of sensor failure and profound knowledge of the accuracy and precision of the sensor data.
- **Trackable/Transferable**—Trackable data can be tracked by specific hardware and software tools conveying information where the data's status is at any point in time. Transferable refers to concepts of method combination applied to different problems.

To meet these SMART monitoring criteria, we suggest an iterative SMART monitoring concept with methods, approaches and tools for *SMART sensors*, *Data-Flow*, *Metadata* and *Sampling* technologies. Figure 6.1 shows this concept with its four overlapping groups of tools.

1. SMART sensor tools enable the interconnection of a large number of sensors, automated data access via standardized and well-documented interfaces, and remote adaptation of measuring schemes to the prevailing measuring conditions. They meet the SMART criteria specific, measurable, robust and trackable.

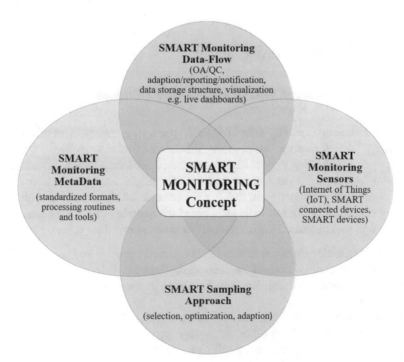

Fig. 6.1 The SMART monitoring concept consisting of four overlapping groups of tools

2. The SMART DataFlow approaches incorporate a variety of standardized data flows within the data lifecycle. Analytical tools and methods provided in automated analysis workflows allow data exploration and analysis to determine a suitable monitoring strategy; this includes methods to classify data, determine outliers, fill data gaps, recognize patterns in data and adapt data to different spatial and temporal resolutions. SMART DataFlow approaches meet the SMART criteria scalable, modular, accepted, adaptive and relevant.

3. SMART MetaData approaches enable a comprehensive description of newly acquired data increasing their reliability. Standardized data descriptions are important for data fusion, joint analysis and interpretation as well as for the creation of training data sets for machine learning. SMART MetaData approaches meet the SMART criteria specific, accepted, relevant and trackable.

4. SMART Sampling approaches. They, for example, support selecting the most representative sampling points based on auxiliary or prior measured data. They meet the SMART criteria scalable, modular, adaptive and transferable.

6.2.2   Pre-Conditions for SMART Monitoring

6.2.2.1   SMART Monitoring and Technological Advancements of Sensors

SMART monitoring of the environment incorporates SMART sensors such as the internet of things (IoT), SMART connected sensors, and smart devices, playing an essential role within SMART monitoring. Sensor technology is improving from a simple measurement sensor to a SMART sensor with local intelligence, decentralized data pre-processing, digital output, and near real-time communication options. Microsystems technology with micro- and nano-electronics low-power computing capabilities, high data volume storage and better batteries are driving this development. An ideal SMART sensor is a sensor in which the complete signal conditioning and signal processing are combined in one unit in addition to the actual measurement acquisition. Such sensors usually include a microprocessor or microcontroller and provide standardized interfaces for communication, e.g. via field bus systems or sensor networks. Thereby, such sensors' complete sophisticated task is to be fulfilled without an external computer to meet miniaturization demands, decentralization, increasing reliability, reducing costs and improved flexibility. The characteristics of a SMART sensor are (Sauerer, 2013):

- provides a digital output signal, often via a standardized interface; in stand-alone systems also via a wireless data connection
- can be addressed via an address and has a bidirectional digital interface
- data transmission and information exchange among different devices and different domains

- executes commands and logical functions (complex measurement value processing up to measurement evaluation) to allow (near) real-time decision-making, service supporting and management
- existence of extensive calibration and diagnostic capabilities
- in self-sufficient systems existence of data memory
- self-sustaining sensor systems with wireless data transmission and energy harvesting eliminate the need for cabling or battery replacement, allowing almost unlimited operating time in hard-to-access locations

Therefore, a SMART sensor is suitable to meet the increasing demands on reliable monitoring tools such as miniaturization, the realization of high data sampling rates with higher accuracy and reliability, acquisition and decentralized real-time processing of spatially distributed measurement data, sensor fusion allowing the combination of different sensor data, ease of integration, (wireless) self-sufficient networking, higher reliability and less maintenance due to self-maintenance opportunities, low power and low latency (data transmission with minimal delay), the possibility of mobile edge computing (MEC), especially for mobile crowdsensing and cost reduction.

A vital facet addresses the monitoring of the sensor function itself. A service reduced and reliable sensor operation, especially in long-term remote-controlled applications, requires modern communication procedures between the sensor and the control unit. Today, even the most straightforward I.T. equipment like printers have fully automatic reconnection, self-analysis, and, if required, also calibration procedures. Any necessary information like driver updates or serial number-related information is available in repositories. The sensor should automatically connect in case of malfunctions or even routinely checks if updates or improvements are available. Unfortunately, this is not the case in most environmental sensors, which often also do not even have the most basic plug-in connection procedures. Therefore, significant technological innovations in sensor development are needed to provide smart monitoring technologies with self-repair mechanisms if the control software fails and reliable alerting functions in the event of contact failure (Fischer 2020). We need to implement state-of-the-art I.T. technology in the field, working based on plug-and-play technology, including fully automated transmission, verification, storage, accessibility of sensor metadata and sensor actions such as deployment or maintenance so that human interaction in sensor operation can be significantly reduced.

6.2.2.2 SMART Monitoring and FAIR Principles

The availability of sensor data represents a prerequisite for parametrization and model validation. Such data fusion and integration demonstrate the importance of implementing FAIR principles for sustainable data management and allows interoperability among different data services (Wilkinson et al., 2016). These principles primarily aim to generate the maximum benefit from data and their metadata by supporting machine-actionable data infrastructure processes making the data findable and accessible by machines and humans. However, it needs to be mentioned that the FAIR principles do not require a detailed description of data quality and do not cover content-related quality aspects. An assessment of data quality and the data provenance is essential to preclude the possibility of inaccurate, incomplete or even unsatisfactory data analysis applying and avoiding poorly derived, misleading or wrong conclusions. For our SMART monitoring approach, the following information needs to be part of any FAIR compliant data:

- available auxiliary or prior measured data
- essential information on smart sensor networks, including timing, ambient conditions and data aggregation issues
- existent, general information regarding both sensor and measurement uncertainties and calibration of sensors in a traceable way
- information on executed processing and analysis steps with their assumptions, e.g. information on quality control/assurance steps or applied Proxy–Transfer Functions to derive parameters of interest
- information on used methods and their assumptions (e.g. test and training data set) of supervised and unsupervised machine learning generated data products
- information on gridding algorithm and their specific assumptions/parameter settings especially for larger scale sensor data that have been converted into a derived data product (grid/raster; correlation)

6.2.2.3 SMART Monitoring and Standardization

As crucial part for joint scientific activities SMART monitoring Data-Flow and SMART Metadata tools will need to support scientific cooperation and data integration through standardized workflows and metadata schemes. Standardization is often emphasized as important process, although there is little awareness about how standardization should be carried out. Standards as expressions of consensus enable in general safety, allow to control processes, increase transferability and also support creativity. Standards can be established by geographical extent or reach (e.g. international and national), by scope, by strength (e.g. regulation versus recommendation) and by subject (e.g. devices, procedures and workflows).

Standards can ensure data reproducibility, a key element of interinstitutional cooperation and joint data analysis. Many scientists mention standardization as the means to make data interoperable, but many are not clear about the requirements for the respective standardization process. Such a process requires a transparent input into the corresponding workflows and daily work to those involved to achieve high acceptance and develop the best possible standard. The smaller the affiliated group, the easier it is to define standardized procedures and apply them within the group/consortium. Therefore, there is a seemingly infinite number of standards for different data workflows which makes a reliable standardization difficult. The Vienna agreement from 1991 allows cooperative standardization efforts between European

Committee for Standardization (CEN) and the International Organization for Standardization (ISO). Many standardization activities are going on in so called technical committees at CEN or ISO. Another faster option is to prepare a CEN Workshop Agreement (CWA) at the European Committee for Standardization (CEN) to reach a broad acceptance internationally (CEN, 2021). This kind of agreement must follow specifically defined steps: preparation, the initiative's announcement, kick-off meeting, drafting, consensus, publication and implementation (CEN, 2021). The advantage of such a standardization effort is that such "lower" standard is being reviewed every 3 or 5 years allowing the consideration of novel developments.

Even with the understanding that standardization is essential within SMART monitoring tools, we could not prepare and start such a CWA process within the Digital Earth project lifetime for our SMART monitoring approach. Such standardization efforts would need to include all steps from data acquisition, data processing and data storage, all described as clear and reproducible as possible. Both, standardized procedures in the data acquisition (monitoring set-up, acquisition, calibration, data cleaning,) and standardized metadata, decide on the usability and trustworthiness of monitoring data. Cooperation with existing initiatives and infrastructures such as DataFlow Framework from Sensor Observations to Archives (O2A at AWI) and Modular Observation Solutions for Earth Systems (MOSES) were intensified to improve existing tools and bundle competencies (UFZ 2021; Koppe et al. 2015; Gerchow et al. 2015).

6.2.2.4   SMART Monitoring and Data Quality

Data quality is a crucial, although not explicitly mentioned, requirement for data FAIRness; it is essential to ensure reusability of data. A documented data quality is required to enable meaningful data selection for data fusion and reuse. Due to the versatile application of low-cost sensors in environmental science, information on data quality has become increasingly important and scientists who acquire and use monitoring data must be aware of the importance of data quality and their trustworthiness. Even though data may be FAIR in terms of availability, the data are not necessarily "good" with regards to accuracy and precision. Unfortunately, there is still considerable confusion in science about what good or trustworthy data are (e.g. Dorgio et al. 2021). Trustworthy data may be achieved by simple/automated data flagging algorithms, ensuring that data are plausible with respect to specific criteria (e.g. threshold). But real trustworthy data imply more when considering accepted standards for scientific data that have an uncertainty value for each measurement. An accepted approach in providing an uncertainty range for single data points/measurements is by providing accuracy and precision as defined in ISO standard 5725–1(1994; accuracy evaluates the proximity of measurement results to an accepted reference value, precision considers repeatability or reproducibility of measurements). This approach allows for a numerical expression of how close a measured value lies with a certain statistical probability to the real value (e.g. 90%).

Even though the uncertainty assessment is commonly accepted and good scientific practice in most experimental studies and measurements, it is not yet a must-have in many monitoring approaches where single sensor data or a time series is often provided without assumptions of the associated uncertainty. Such assessments are of significant importance when classifying data as relevant or good for a specific application or scientific question. Discriminating, e.g. different water masses based on temperature and salinity requires high accuracy and precision. Estimating the presence or absence of a specific fish species based on the concept of fundamental and realized ecological niches using the same two parameters, temperature and salinity, allows for a much larger accuracy and precision range of the data. Therefore, the same data cannot be used by e.g. an oceanographer but by the behavioural ecologist. Each scientist must be enabled to decide if data are good or have to be rejected as probably unfit for a specific scientific question; without an assumption of the data uncertainty, this is practically impossible.

Modern data science methods such as machine learning can blend diverse datasets even with lower quality to gain valuable information. However, to assess and interpret such results, knowledge of data quality is required. Traditional repositories hosting data from scientific or regulatory monitoring initiatives as well as from scientific field campaigns could usually rely on more or less rigid quality assurance chains. According to the international organization ASQ, quality assurance can be defined as "part of quality management focused on providing confidence that quality requirements will be fulfilled." Quality control as the "part of quality management focused on fulfilling quality requirements" is essentially the inspection component of quality assurance (ASQ, 2021). Quality control (QC) is distributed across data acquisition, data management and data curation tasks and should be discussed as such and jointly executed by the involved people scientist check the validity of the data values themselves whereas data manager and data curators possibly focus more on metadata quality (completeness, standardized terminology etc.). Similar quality control issues occur in different monitoring domains, scientific disciplines and involve many different data types. Recently, many projects and initiatives have begun to harmonize data quality control efforts (e.g. ENVRI-FAIR and NFDI) and develop software tools to assist in quality control across various environmental research domains (Schultz et al., 2019; see also Sect. 6.4.1 for examples). Such initiatives may only affect the data processing steps but have also considerable effects on entire monitoring set-ups.

6.2.3 Future Tasks to Further Increase Smart Monitoring Efforts

The development of new and faster machine learning and generally AI tools will undoubtedly bring new possibilities for advancing tools of SMART monitoring. Despite the integration of these new methods we see a number of essential tasks as important for future applications of SMART monitoring (Table 6.1). These must go

Table 6.1 Future needs and challenges for SMART Monitoring Tools

SMART Sensors	• fully automated transmission, verification, storage, accessibility of sensor data and their metadata • automated sensor maintenance and set-up • powerful synchronization and data management system for wireless sensor networks (real-time clock; data versioning, …)
SMART DataFlow	• standardized approaches, e.g. documentation/reporting, QA/QC, data gap filling, statistical analysis, quality assessment and visualization • quality assurance routines and metadata description for supervised and unsupervised machine learning algorithm to achieve reproducibility of the results • near real-time visualization with incorporated analysis tools
SMART MetaData	• agreement on a shared vocabulary for the metadata elements and values as an interdisciplinary approach • automatic filling of metadata (e.g. electronic and paper documents, software) and • automatic error checking to reduce human errors automating metadata management
SMART Sampling	• automatic grabbing of all available web data for a selected area with various resolution to save and visualize this data in a standardized format • automated bundling of all available data (former monitoring data, satellite data, auxiliary data) from a selected area and • determine the representative sampling area or points accordingly

hand in hand with additional IT security when distributed dataFlows and IoT sensors and data repositories become the new standard for monitoring the Earth environment.

6.3 SMART Monitoring approaches and tools

In Digital Earth, we developed several approaches and tools to enable the expanded SMART monitoring concept we set up in Digital Earth.

6.3.1 Hard-and Software Tools for a Modern Communication between Sensor and Control to Enhance Traditional Monitoring Efforts

Environmental research is changing towards using monitoring strategies that are no longer based on static data collection, but on the coupling of prediction and empirical data that integrate sensors near real-time data stream for continuous modelling. The challenge, that in practice, requires a sophisticated implementation of a decision and control basis at sensor level. While sensors used to be rather one-dimensional and stupid data suppliers, nowadays complex sensor systems where several sensors are interconnected are being used increasingly. An important aspect in this respect is the implementation of a data model according to a holistic data processing e.g. the Lambda architecture (e.g. Kiran et al., 2015). In this way, sensors do not just deliver measured values but entire message packages (e.g. JSON) via defined protocols (e.g. MQTT) and interfaces (e.g. HTTP). In addition to the measured value, these message packages contain various descriptive metadata about the sensor and the situation of the measurement.

This procedure allows complex algorithms and analyses to be applied directly to the data stream based on the message packets. Thus, the approach in applied environmental monitoring shifts from pure data collection to adaptive measurement in the context of an application or a specific scenario. The measured sensor value is embedded in a context and thus receives a clear space–time reference as well as a context-related allocation. The fusion even of heterogeneous data streams is thus considerably simplified, since connecting descriptive parameters are available within the metadata, which allow linking different message packets.

The main innovation of the process flow is that data collected in the course of monitoring can be directly related to a-priori information. It is irrelevant whether the context is based on modelling or accompanying measurements. Since the infrastructure and the underlying data model represent an always existing and complete solution space, the monitoring efforts can be constantly optimized, similar to a machine learning approach, because the set of rules for data sampling/sensor measurement (sampling interval, additional sensors on/off, ...) are subject to constant proving. This approach allows a broader or rather holistic assessment of varying, large-scale environmental phenomena. To do so, there is a corresponding need for capable hardware and software tools that are specialized to execute such an assessment in a tailored way.

Figure 6.2 gives an illustration of a data stream architecture with real-time data processing and IoT-capable sensors. Starting with the sensor technology, sensors must not only transfer the results of the measurement conversion (e.g. calculating turbidity from a voltage signal of a turbidity sensor based on light backscatter) but obviously also needs to provide information about the context (e.g. calibration, application conditions). In the next step, a gateway is needed to collect the sensor data and harmonize them according to the data model and assign them to a reference system with the information about place, time and sensor ID. The gateway also serves to define global parameters such as the sampling rate or to ensure time synchronization. The time base is the foremost quality assurance criterion for the implementation of such a sampling paradigm.

Once the message packets are in the data stream, the downstream processing steps are borrowed from other domains such as logistics or business informatics. Low-code programming for event-driven applications can be achieved, for example, by using Node-Red as a powerful and versatile platform (https://nodered.org/) for connecting hardware devices, APIs and online services. Time series databases have proven to be an efficient and robust solution for storing sensor data, e.g. influx DB (https://www.influxdata.com/). Thanks to their own syntax and robust architecture, the query and storage of even large amounts of data is very fast and allows establishing complex

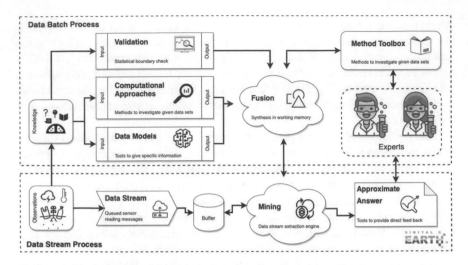

Fig. 6.2 Data stream architecture for processing real-time data based on IoT-capable sensor systems and considering server-side services for the valorization and contextualization of environmental monitoring data

query and processing procedures. As such, a data-driven architecture for service-oriented observation methods and in-stream process modelling close to real time is ready for use, available as open-source with powerful capabilities thanks to a large user community. This makes such an approach rather low-cost with low overhead for not directly necessary tasks.

6.3.2 SMART DataFlow

A SMART DataFlow from the sensor to the database is a central part of the SMART monitoring concept as highlighted in Chapter 6. This DataFlow represents parts of the data life cycle and plays an important role in data acquisition, data handling and data management (Fig. 6.3). Within a SMART DataFlow standardized and automated procedures are needed to obtain reliable data for subsequent analysis and application.

To gain reliable and trustworthy data, it is essential to develop and apply standard operating procedures within the DataFlow from the individual sensor to the repository. The following prerequisites or conditions have been identified as important:

- Availability of a portfolio of different independent methods measuring a specific parameter
- Combination of various SMART sensors in a network allowing self-repair / calibration mechanisms during operation

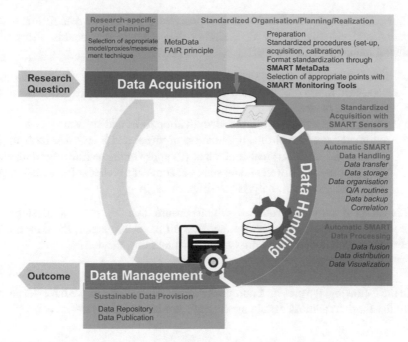

Fig. 6.3 SMART DataFlow as backbone of the SMART Monitoring concept

- Possibility of applying a hierarchical monitoring approach across different scales and parameters
- Possibility to transfer concepts of combined methods to further research questions
- Capacities for automatic data handling using standardized data transfer, data storage, Q/A routines and data backup rules and routines in, e.g. traditional repositories; important are:

o defined data or formats and standardized data format transformation proce-dures

p standardized routines for Q/A to ensure identical data flagging; standardized routines should correspond to existing internationally accepted and applied Standard Operating Procedures (SOPs), e.g. from ICOS, ARGO, the World Meteorological Organisation Global Atmosphere Watch programme or more informal at GO-SHIP from the marine science (https://www.go-ship.org/Hyd roMan.html)

q standardized, automated and interactive uncertainty analysis tools for data and proxy transfer functions

r standardized processing routines for integrated data analysis or proxy transfer functions

- Capacities for automated SMART data processing that allows data fusion, distribution and visualization of all available data such as dashboards. Processing should include:

    o   hierarchical data storage structure to combine and integrate relevant auxiliary data, calibration protocols and data from intercomparison experiments of the respective device

    p   standardized data analysis and visualization tools and software

    q   automated (near) real-time application of visual and interactive tools to: (1) process various (near) real-time data; (2) apply multiple filters for data analysis; (3) combine different data sets; (4) connect to other software packages; (4) enable joint data analysis by different users

- Standardized reporting routines which ensure that all processing steps are precisely described and traceable and that all users can assess the data quality of the parameters derived by proxy transfer functions, needed are:

    o   standardized metadata vocabulary and schema

In the following, we present a number of approaches and tools we have developed in Digital Earth to address certain aspects of the SMART DataFlow.

6.3.2.1 Automated QA/QC pipelines (Quality Assurance/Quality Check)

During the Digital Earth project, automated workflows for data processing have been developed, focusing on the near real-time quality assessment and quality control of the collected data. One major design criterion for the workflows was their composability with existing workflows of other users and their scalability to be easily adaptable to other requirements. In this subchapter, two examples of the successful implementation of the developed workflows in the European Research Infrastructures IAGOS and TERENO are introduced

IAGOS: The European Research Infrastructure IAGOS (In-service Aircraft for a Global Observing System; www.IAOGS.org) operates a global-scale observing system for atmospheric composition and essential climate variables by deploying automated instruments on passenger aircraft during their commercial flights. To handle the immense DataFlow from the fleet of aircraft collecting data, IAGOS has implemented an automatic workflow for data management, organizing the DataFlow starting at the sensor towards the central data portal located in Toulouse, France. The workflow is realized and documented using the web-based Django framework with a model-based approach using Python (Fig. 6.4). In Fig. 6.1, the overall sketch is shown.

A permanent active cronjob called Task Manager (outer box) activates an individual task instance (dotted box) of a task class describing the complete data handling process. This includes the following steps: (1) The Transfer Handler checks for new

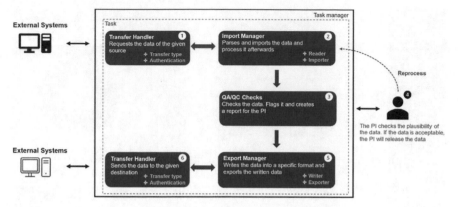

Fig. 6.4 *Scheme of the IAGOS process chain including a QA/QC pipeline*

data of the task-specific data type by using a RESTful API (Application Programming Interface) operated at the data centre in Toulouse. The API handles the necessary authentication by using an individual ssh256 encrypted token generated with a pre-shared passphrase and unique token (timestamp).

If new data is transferred, the data is passed to the Import Manager (2). The Import Manager reads and parses the raw files (pandas toolset) and processes the raw data to meaningful values. In the end, the Import Manager stores the processed time series to the instrument database for further processing. As the next step (3), the advanced QA/QC Handler performs checks, flags the data and produces a report for the PI who has to release the data for Level 1 and Level 2. (see Table 6.2).

In principle, step 2 and 3 are the same for all different data levels. In the end, the data-level reached depends only on predefined requirements e.g. the availability of the post-flight calibration. In step 4, the Export Manager writes the data to a specific transport format (e.g. NetCDF, or API specific format) and passes it to the Transfer Handler (6), which finally handles the transfer for a specific data type towards the data centre, including the authentication process already described for the Transfer Handler (1). The Task Manager tracks the status of all tasks even if they are terminated by reaching Level 2 or stopped by the PI. All information on tasks, including decisions of the PI, is stored for later reprocessing if needed.

Table 6.2 Definitions of the level of maturity for IAGOS data

Level 0	Raw data checking
NRT	Fully automated upload of processed raw data including automated QA/QC checks and flagging within 3 days (Near Real Time)
Level 1	NRT grade data, approved by PI
Level 2	Fully reprocessed scientific grade data, including post flight calibration, automated QA/QC and flagging approved by PI

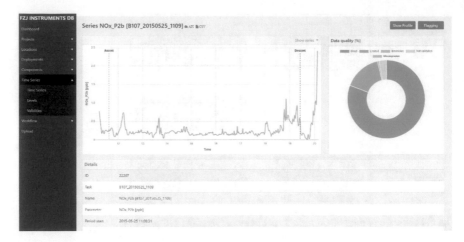

Fig. 6.5 Screenshot of the IAGOS Instrument Database application for visual quality check

This workflow performs all necessary data processing and QA/QC tests to automated upload NRT processed data and serves the PI as a basis for approval decisions. This includes repeated cycles for different stages of data maturity. The PI can monitor the status of all tasks through web-based reports produced by the Task Manager. An automated reprocessing is possible by storing metadata on all steps as well as decisions of the PI. Implementing the workflow is one big step to making IAGOS data handling compliant with the FAIR principles (Fig. 6.5).

The automated QA/QC tests are accessed inside the workflow using the Python framework **Autom8QC** developed and used by the DE community. It fulfils the following prerequisites:

- Application of probabilistic approaches
- Easy adaptation of test strategies
- Adaptation to different environments
- Application of different test measures (e.g. logical-, statistical-measures)
- Combination of tests in groups
- Combination of tests in sequences

The following QA/QC tests are already implemented for common use (Table 6.3):

The described framework was successfully integrated into the IAGOS workflow to create a QA/QC pipeline that generates an automated test report and automatically flags the measured and processed data. The PI uses these reports for the final approval decision (data maturity Level 1 and Level 2).

TERENO is an initiative funded by the Helmholtz Association to create observation platforms to facilitate the investigation of the consequences of global change on terrestrial ecosystems and the socioeconomic implications of these (Zacharias et al., 2011). Four observatories have been set up in 2008, each operated by one

Table 6.3 List of implemented test modules for the Auto8QC framework

Test type	Test name	Description
Value	Flatline test	Identifies repeated occurrence of one value in a time series
	NaN test	Test for "Not a Number" values
	Specific Value test	Test for a specific value
Time	ITT	Increasing time test (monotonic timeline)
	Time Gap test	Test for unexpected gaps in the timeline
	Time Range test	Test for specific timeline range
Limit	Global Minimum test	Test for values below specific minimum
	Global Maximum test	Test for values above specific maximum
	Global Range test	Test for values outside of specific range
Outlier	ESD (extreme Studentized deviate) test	Detects one or more outliers in a univariate data set
	LOF (Local Outlier Factor) test	Compares the density of any given data point to the density of its neighbours
	IQR (Inter-Quartile Range) test	Detects outliers using Inter-Quartile Range
	MAD (Median Absolute Deviation) test	Detects outliers using Median Absolute Deviation
	OutlierZ test	Detects outliers using the Z-score
Peak	ScipyPeak test	Uses the Python library SciPy to detect peaks

Helmholtz Centre, which maintains its local data infrastructure. The individual infrastructures are interconnected into the distributed TERENO Online Data RepOsitORy (TEODOOR), supporting the acquisition, provision, and management of observations via SWE specifications and several other OGC web services (Kunkel et al., 2013).

For the Eifel observatory, operated by the Research Centre Jülich, about 180 mio. mostly meteorological, aquatic and terrestrial observations are collected each year, from which about 90 Mio. (54%) have to be quality checked. Each observation is, among others, attributed to a processing status and a data quality flag. Following the IOC of UNESCO (2013), we adopted a two-level scheme to assign the data quality for each observation. The first level defines the generic data quality flags, while the second-level complements the first level by providing the justification for the quality flags based on validation tests and data processing history. In TERENO, the second-level flags are specified by the domain experts. The processing status describes the data type and determines the workflow for data editing and publication (Table 6.4).

Observation data are imported into the observational database and managed with our time series management system (TSM 2.0) (Kunkel et al., 2013). The system includes a highly configurable file parser, a data processor as well as a task manager

Table 6.4 Data processing levels for TERENO data

Level	Descriptions	Data Source	QC	Data Editing	Availability
Level 1	Raw data	Automatic importing or manual upload	No	Not allowed	Internal (on request)
Level 2a	Externally quality controlled data; an approval is pending	Level 1 data (manual upload)	Yes	Not allowed, flagging only (except human observations)	Internal (on request)
Level 2b	Quality controlled data with automatic QC procedures	Level 1 data (automatic upload)	Yes	Not allowed, flagging only	Internal (on request)
Level 2c	Externally quality controlled data with an expert approval	Level 2a data	Yes	Not allowed, flagging only	Public
Level 2d	Quality controlled data with semi-automatic QC procedures (automatically and by human)	Level 2b data	Yes	Not allowed, flagging only	Public
Level 3	Derived data	One or more Level 2 data	Yes	Allowed	Public

for internal and external procedures. It allows automated data pre-validation such as transmission and threshold checks and flagging of the data along with the importing process. After a visual examination by the responsible scientist or technician, data of Level 2 or higher are made available online automatically via Sensor Observation Services, which provide the data, the processing levels and the data quality flags (Devaraju et al., 2015). The full process of data collection, transmission, processing, QA/QC and management is fully documented and certified according to ISO 9001.

However, several issues and limitations arise with this QA/QC workflow, like:

- QA/QC is performed inside the TSM during the data import with the advantage of very fast processing of the data under consideration of parameters and sensors. In practice, however, the system is limited to basic QA/QC routines like threshold checks.
- Implementing additional, more complex and/or site-specific QA/QC routines and its application to specific data processing workflows requires significant programming skills, making it almost impossible for scientists to develop these routines by themselves.
- For these reasons, the scientists will usually download the data to develop and perform their own QA/QC routines on their computer systems. In most cases, processed and/or QA flagged data will not get back into the infrastructure.

To overcome these limitations, we extended the workflows within the DE project for automated data processing by Auto-QA. This browser-based flow editor that allows users to develop, test and run their own processing chains to add their own procedures to access the data infrastructure using standardized interfaces and to run, visualize and upload the results. As a basis, we used the Node-Red software, a flow-based development tool for visual programming. Node-RED, built on Node.js, provides a web browser-based flow editor, which can be used to create JavaScript functions. Elements of applications can be saved or shared for reuse using JSON. Red-Node uses "nodes", which can be interconnected graphically to processing chains. Each node can be characterized by individual properties, which can be used for process control.

In order to use Node-Red for TERENO, we developed a library of modules for importing, visualization and exporting data and included the QA/QC routines from the Python framework Autom8QC. The framework was extended by a module for automated conversion of PYTHON code into JAVASCRIPT to allow the incorporation of custom PYTHON modules in Auto-QA.

Figure 6.6 shows a simple example of a global range test of an observed parameter. The process will be initiated by a trigger, which may be a manual start or, for instance, a GET request via HTTP to this particular trigger. The trigger initiates the data to be read from TEODOOR by an OGC-SOS client (CLISOS). The output produced by this node will be sent to the variables and min_max node. "variables" will then print all available variables to the debug panel. "min_max", which is a global range filter, flags the data according to the settings made in the node. The output will be formatted in the "clisos_format" and then be outputted to the debug panel. The data sent to the debug_panel can be visualized graphically, written to a "file" node or uploaded directly into the infrastructure using a SOS-T. The whole process can be parametrized by the GET parameters of the trigger and stored.

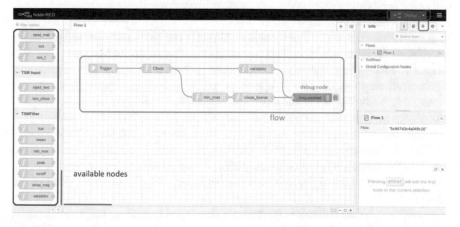

Fig. 6.6 Screenshot of a simple QA/QC workflow for a simple global range test of an observed parameter

This makes is possible to use the developed workflows as templates for the application to other parameters and/or sites. Moreover, the Auto-QA can easily be initiated automatically through the TSM, by calling the trigger through the CRON module of the TSM followed by an automated import of the flagged and/or produced data. Finally, Auto-QA is not exclusively used for QA/QC, but also for the development of workflows for extended data processing, e.g. calculating runoff data from water levels on surface water gauging stations or the calculation of soil moisture data from Cosmic Ray stations.

6.3.2.2 Analytics of Big Geo-Social Data for Environmental Monitoring: Exploring Microblogs to Spatiotemporally Characterize Floods, Droughts and Typhoons in China

In order to better understand and respond to the occurrence and impacts of extreme hydrological events, SMART monitoring should also integrate data and metadata related to citizens' perceptions of floods and droughts.

As the trend towards social media data has increased rapidly over the last decade due to the simplicity and accessibility of social media platforms, it is obvious to develop a web scraper and data filtering tool for an environmental-hazards-analysis as well, to automatically filter out their spatially and temporally distributed occurrence from microblogging services such as Twitter. Citizen Science could thus serve, for example, as first-hand information on the hydrological situation in communities.

Although more and more remote sensing data are being used to monitor hydrologic events in support of traditional hydrologic monitoring networks, observation is limited by the coverage of monitoring networks and the spatiotemporal resolution of satellite imagery, and thus its informative value. Alternatively, the highly diverse and rich data from social networks contain valuable meta-information on hydrological events and can be extracted using keyword-based sentiment analysis techniques (Olteanu et al., 2015; Wang et al., 2016).

An automated approach that focuses on China in our study retrieves and processes microblogs from Sina Weibo. This approach consists of an API-independent web scraper to retrieve large volumes of microblogs, a data cleaning and filtering module, a georeferencing module and a supervised machine learning approach to classify content reliability (Fig. 6.7). The workflow was applied to analyze more than 700,000 microblogs on typhoons, droughts, and floods from 2018 to 2019. For validating the extracted information, remote sensing data from International Best Track Archive for Climate Stewardship (IBTrACS) (Knapp et al., 2010 and 2018), standardized precipitation evapotranspiration index (SPEI) drought index (Vicente-Serrano et al., 2010) and NASA Global Precipitation Measurement (Huffmann et al., 2014) were utilized.

In addition, a data collection system was developed to capture large social media data from Sina Weibo, a popular microblogging website in China. Microblog details, including content, authors, time of publication and geotags, were captured by a Python script that used packages to log in and parse HTML scripts from websites

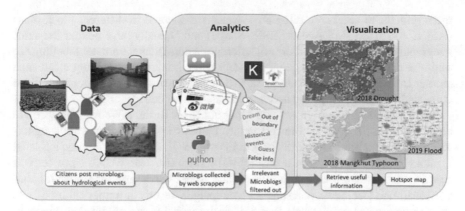

Fig. 6.7 Simplified process chain for utilizing microblogs to predict floods

to capture targeted data. Keywords related to floods, droughts, and typhoons were prepared and sent as search queries to the Sina Weibo search engine, and the collected data were stored.

In the analysis of microblogs related to typhoons, the method works qualitatively very well and the spatiotemporal distribution of microblogs reflects the actual course of typhoons. This makes this type of workflow a potential tool for tracking the trajectory of meteorological events with sufficient social attention. In the case of droughts, where our study is one of the first researches on this topic on Sina Weibo, about three-quarters of the microblogs are located in areas classified as dry according to the SPEI drought index, which shows that this approach could provide meaningful information here as well. Interestingly, however, limitations were encountered in the analysis of flooding, as the distribution of microblogs was highly dependent on population density, and furthermore could not be correlated with precipitation patterns and river flows over a wide area. Future improvements can be made by including additional social media data sources to achieve even higher data density.

In conclusion, while not all microblogs were useful for our data analysis, and their content needs to be better cleaned for analysis (e.g. removal of assumptions, past events, advertisements, and reposts), the integration of social media into monitoring approaches is a big step towards more data-based environmental research.

6.3.2.3 Establishing a Webgis Project for Hydrological Campaign Planning and Data Sharing at the Mueglitz River Basin

Planning event-driven monitoring campaigns on spatial catchment scale requires a comprehensive overview of existing measurement/monitoring locations, previous campaigns, and the distribution of hydrological, geological, and geomorphological features within the study area. In Germany, data and related meta-data are often distributed among various institutions at the local, state and federal level, making them difficult to access. Additionally, datasets are stored in different file formats (ASCII, MS Excel, MS Access, ESRI Shape) and typically have neither the same geographic coordinate system or projection nor a consistent parameter labelling. To optimize data access for campaign planning while minimizing working efforts and storage capacities (several data requests to external data providers), better centralized data handling, processing, and access approaches are needed.

For the MOSES intensive test site Mueglitz River Basin, a WebGIS project was established via the AWI GIS infrastructure maps.awi.de. The framework's significant components are an ArcGIS Server, a PostgreSQL database including a Spatial Database Engine (SDE), and a desktop GIS software. The WebGIS project was based on datasets from state authorities (Sächsisches Landesamt für Umwelt, Landwirtschaft und Geologie), measurement campaigns, results of numerical models as well as environmental datasets that are freely available via online data repositories. The links to the original datasets can be found within the AWI WebGIS. Dataset projections, formats and metadata descriptions have been standardized in close cooperation between the project partners following ISO standards. The data and metadata provided by the WebGIS include:

- Locations of weather stations, monitoring gauges and drilling logs
- Modelling results from the OpenGeoSys groundwater flow model (Ver. 5.7; https://www.opengeosys.org/ogs-5/) and the DIFGA rainfall-runoff model (Schwarze et al., 1991)
- Maps of soil, geology, rainfall and land use patterns in the study area
- Locations of sampling and installations during the MOSES campaign 2019

Rich metadata information is given for the "MOSES/Digital Earth Müglitz Campaign" project on https://maps.awi.de/awimaps/catalog/ (Fig. 6.8). The current collection of project data on maps.awi.de is not only for scientific purposes, but it is also open to the public and enables the knowledge transfer to the non-scientific community. All data are provided through Open Geospatial Consortium (OGC) standardized Web Map Services (WMS) or Web Feature Service (WFS), which facilitate data exchange and data visualization among project partners. Based on the experience of the joint-work during related MOSES campaigns, the framework of such a WebGIS project will support upcoming campaigns and serves as a good blueprint for an "almost seamless" DataFlow for campaign planning.

6.3.3   SMART MetaData: Without Trustworthy Descriptions, Data can be Un-FAIR

Documented information about why and how data were collected, their structure, quality assurance, confidentiality, access possibilities and terms of use are typical metadata information. The identification and tracking of different datasets versions are essential in order to find, use, share and manage data especially over longer

Fig. 6.8 Screenshot of the Müglitz campaing at the maps.awi webpage

time in a sustained way. Metadata thus describe other data or information about an object, be it physical or digital, to facilitate search, evaluation, acquisition and use of resources (Duval 2001). Metadata can be descriptive and consist of information about the data content and context (e.g. title, keywords, abstract); they can be structural provide information on the relationship between different datasets and they can be administrative which is essential to manage the data with respect to e.g. ownership and rights management. Accurate and complete metadata are a prerequisite for data sharing and interoperability across different data types. However, the process of describing and documenting scientific data has remained a tedious, manual process even when data collection is fully automated. Researchers are often reluctant to share data with good metadata information even with close colleagues because creating documentation takes much effort and time.

In SMART monitoring, the availability and incorporation of metadata and auxiliary environmental data play an important role for data and knowledge improvement. Below we show an example of how good metadata information can be structured and what might happen if metadata description is lacking and data become not usable anymore.

The COVID-19 pandemic shows the importance of mathematical models for understanding the spread of the disease which is at the base for deriving mitigation measures (Bjørnstad et al. 2020; Dehning et al. 2020). In this process, it is essential to determine and validate relevant model parameters and data integration from different sources is a prerequisite for parameterization and validation of predictive tools or models. The needed data integration calls for having FAIR principles in place for a reliable analysis and interpretation of the data (Wilkinson et al. 2016).

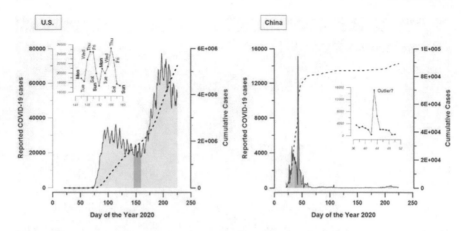

Fig. 6.9 *Daily and cumulative cases for the U.S. with an impressive example of a weekly periodicity and for China with an example of a rapid increase of daily numbers due to changes in counting* (Dong et al. 2020)

COVID-19 data illustrate the importance of data description with respect to origin and data quality to interpret data accurately. Figure 6.9 shows an example of reported daily COVID-19 cases for two different countries, here the US and China. Visual analysis of the curve of the daily cases reveals an approximately seven-day overlap of a slow trend process and a higher frequency process of weekly periodicity. Today, we know that the higher frequency process represents an artefact of data acquisition as a result of that laboratories and authorities that collect the board's data do not work on weekends. The behaviour of such an artefact can change over time if the data acquisition and communication processes changes/adjusts over time. A specific kind of adjustment is a change in counting as visible in the curve provided for China. The evaluation of the total number of covid cases with such sudden changes could lead to doubts concerning the completeness of the data.

An assessment of data quality and data origin is essential to preclude the possibility of inaccurate, incomplete or even unsatisfactory data analysis particularly when automated methods are applied that may lead to misleading or incorrect conclusions. The term "trustworthiness" of data summarizes all the aspects related to this potential problem. The fundamental features of trustworthiness are validity, provenience/provenance and reliability (Fig. 6.10). Using secondary data demand an additional detailed description and assessment of their reliability and validity which causes an increase of data collection methods. Validity assessments apply defined procedures to check for the accuracy of the observation findings. Data reliability analyses evaluate the quality of research by indicating the observed data's consistency and stability (repeatability and reproducibility).

The data reliability evaluation, especially in environmental science, should assess how precise and accurate individual measurements and with which uncertainty the measured value reflects the real value. It is evident that all acquired data come with an inherent uncertainty (Paasche et al., 2020). It is close to impossible to measure any

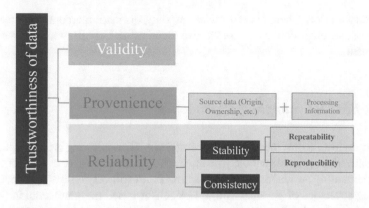

Fig. 6.10 Components of data trustworthiness, which should be considered to avoid wrong analysis and incorrect conclusions. A detailed description of the topic can be found in Koedel et al. (2022)

environmental variable with 100% certainty due to numerous limitations such as non-representative sampling schemes, improper sample handling, limited expertise of the data collector, unspecified effects of environmental conditions on the observed data, effects of specific electrical components of the measuring device on the data or data collector's bias. Even if this uncertainty cannot be assessed complete a description of the "where, how, who" provides valuable information for further data analysis.

In the case of COVID-19 example, the information about the overall test number to a country's population, delivery time to laboratories, quality aspects of the test centres (experience, analytics devices, number of confirmed invalid or incorrect tests), the daily processing capacity of laboratories, laboratory operating hours, reporting lag time and known test uncertainties is all important to understand and interpret the data in a consistent and comparable way.

Especially for these data and the derived measures, a comprehensive data analysis, including the analysis of stationary trends (7-day trend) and possible anomalies and uncertainties, is required to maintain the population's support. For all kinds of data collection, international efforts should be made to assess the data reliability in a standardized way.

Data provenience and provenance supply important information on the data source (provenience) and the applied processing steps (provenance), e.g. in the laboratory, to understand the data and its trustworthiness. For the given example it makes a difference if the data come from a certified laboratory that has consistently followed standardized routines or not.

It should be all scientists' task to request essential information on data quality from the data provider and to support the authorities through internationally accepted standardized workflows and metadata schemes. However, for most disciplines and their observations there are not yet (universally) accepted standards such as Standard Operating Procedures or suitable representation of uncertainty or other of indicators of validity. All this detailed information on data trustworthiness allows scientists to find appropriate tools and methods for FAIR data handling and a more accurate data interpretation. There is need for an ongoing discussion among scientists, data managers and other data users to establish standardized criteria for trustworthiness assessment.

6.3.4   SMART Sampling Approaches

Another important development concerning SMART monitoring is the adaptation of statistical and AI-related methods to evaluate the location and time of data acquisition for implementing an adaptive sampling approach. A still very typical approach is that sampling strategies of physical samples or sensor-based measurements in the field are pre-determined following e.g. a random or targeted design (sampling at dedicated locations of interest) or a temporal- and/or spatial grid-like sampling scheme. Applying a SMART sampling approach means a pre-designed sampling strategy is adapted based on the specific and additional/external continuously accumulating data and knowledge during the sampling procedure itself.

When trying, e.g. to locate and quantitatively sample a plume of any substance in a specific volume (typically the ocean or atmosphere), it will be SMART to adapt sampling locations and sampling interval and density based on the measured concentration gradients during the sampling itself. Following an arbitrary but regular grid covering the assumed volume the substance might disperse in, can result in incomplete data. Adaptive sampling requires both auxiliary and real-time data that combined with advanced statistical or AI supported procedures and algorithms define better sampling locations or times for a specific monitoring task. Such an advancement in monitoring will not only support decisions about sensor locations but also sensor settings and the monitor strategy in time and space in an iterative way. For this kind of SMART Sampling approach additional supporting tool need to consider other essential data like: previously measured data in the area, auxiliary data for characterizing that area (land use, geology) potentially from remote sensing data, real-time sensor measurements if exiting and ideally modelled data. The aim is to apply mathematical and statistical methods and tools for deciding on where, when and how often sampling should happen, and how reliable sampling points, correlation functions and interrelationships of processes can be derived.

The resolution of the result depends mainly on the input parameters. The resolution of the inhomogeneous data input (remote sensing data, soil map data, land use data,) must be identical to run the clustering algorithm and the output only occurs on the common, coarsest grid size.

Therefore, a cascading coupling of the algorithm should be aimed at. For example, remote sensing data, representing data for large areas, often have a 10–50 km resolution and e.g. a sensing depth of 0.01–0.05 m for soil moisture depending on the measurement principle, frequency, and polarization direction. Within these data representative areas can be determined but downscaling of such data is possible but comes with an increase uncertainty during analysis and interpretation. Mesoscale data (10 to 1000 m) such as geophysical or hydrological data can be used to determine representative sampling points in areas of interest and/or areas derived from clustering of remote sensing or model data. If predictions on representative depths are necessary, as for soil moisture, direct sampling data or other depth-oriented measurement methods (e.g. direct push methods; Dietrich and Leven, 2006) must be included. At which step model data are added depends on the resolution and representatives of the modelled data. Three studies that aim at determining sampling points, extrapolate metal resources in the deep sea and predict organic carbon deposition in the oceans give examples of typical SMART Sampling applications as part of SMART monitoring efforts.

6.3.4.1 How to Determine Representative Sampling Points at Meso-Scale

The determination of representative sampling points regarding the scientific question and regarding already existing data is a prerequisite to save time and workforce and to collect the best suitable data. The challenge is to achieve this in easy, quick and reliable manner, even in the field for a quasi-instant use.

Machine learning tools can be applied to determine representative sampling points in a specific area of interest. For example, clustering mechanisms, as the Fuzzy C means (FCM) algorithm or weighted conditioned Latin Hypercube Sampling are used for exploring multidimensional data with no prior knowledge of possible data relations (Hoeppner et al. 2000, Paasche et al. 2006). Therefore, these methods are applied to identify representative areas for sampling within the much larger area of interest. The application of Fuzzy C-Means Clustering algorithm by Paasche & Tronicke (2007) allows to identify areas of common features and to cluster multidimensional data by assigning each point a membership in each cluster centre. FCM is based on iteratively minimizing an object function for a defined number of clusters and provides the optimum locations of the cluster centres and the degree of partial membership of the clustered data points to the clusters (Paasche et al. 2006). The normalized classification entropy (NCE) indicates the optimum number of clusters by analyzing the membership distribution (Paasche et al. 2010). Previous experiments showed that fuzzy c-means (FCM) with spherically shaped clusters and Gustafson-Kessel clustering (GK) provides good clustering results. However, land use data as categorical variables represent a challenge in implementing this cluster algorithm. Heterogeneous data of this type are not directly usable to clustering methods. First, grouping in larger groups was necessary. Then, the Gowers generalized coefficient of dissimilarity was applied to calculate the distances with L1 (city block) (Gower, 1971; Gower & Legendre, 1986).

The area around Dittersdorf in the Müglitz river catchment area is a focus area of the MOSES hydrological extremes campaigns and as part of this, well-informed sampling locations were needed. To identify areas of common features, the Fuzzy C-Means clustering algorithm was applied to two data sets describing large areas with reasonable low resolution. These datasets were (1) digital elevation model, slope and

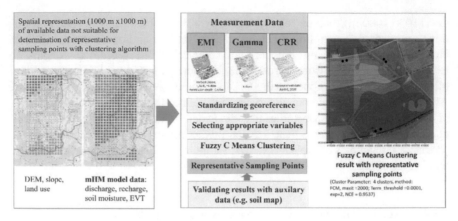

Fig. 6.11 Cascading coupling of clustering algorithm with selection of optimal sampling points at a medium scale

land use (as a categorical variable), and (2) mHM model data such as discharge, recharge, soil moisture and evapotranspiration from 1950–2009 (Fig. 6.11). Spatial representations of these clustering results are not suitable for the determination of representative sampling points because the grid-scale was ~ 1000 m x ~ 1000 m for all input data. Therefore, two meso-scale geophysical data (gamma ray, electromagnetic induction) and cosmic ray measurements were used to determine representative sampling points in Dittersdorf. Gamma ray measurements detect the decay rates of radionuclides with long decay times in soil using a scintillation detector with single sodium iodide crystals to determine the potassium, thorium and uranium concentrations and the natural gamma dose rate. The electromagnetic induction measurement is a highly adaptable none-invasive technique that measures the apparent bulk electrical conductivity of soil (ECa) to get information about field heterogeneity of soil texture and soil water content (Schmidhalter et al., 2008, Viscarra Rossel et al., 2011). Soil moisture content on a horizontal scale of hectometers and at depths of decimeters can be inferred from measurements of low-energy cosmic ray neutrons that are generated within the soil, moderated mainly by hydrogen atoms, and diffused back to the atmosphere (Zreda et al., 2012). Cosmic Ray Neutron Sensing's(CRNS) mobile application is a promising approach to measure field soil moisture noninvasively by surveying large regions with a ground-based vehicle (Schrön et al. 2018).

After standardization of existing georeferenced measurements, appropriate variables were defined and applied to the Fuzzy C-Means Clustering Algorithm. Finally, representative sampling points could be chosen for the meso-scale area. Such a cascading clustering allows the application of heterogeneous data scales and selecting representative areas or points at different scales.

Such a cascading approach can also be called a hierarchical approach. A combination of the most representative areas and then most representative sampling points and further most representative sampling depth allows an effective monitoring approach and saves costs and workforce. Eventually, the provision and use of additional category variables such as passability or entry permits in the cluster algorithm also improve the adaptive monitoring and sampling approach.

6.3.4.2 Using Machine Learning For Automated Site Detection of Massive Seafloor Sulphides

Many current research questions in marine sciences are related to understanding the complex processes that govern resource occurrences. Relationships between the driving forces and response functions are complex, multi-faceted and usually non-linear. Additionally, multivariate and multi-disciplinary data acquired in the marine realm span disparate spatial scales and cross traditional geoscientific domains like geophysics, geochemistry or geology. Integrative data assessment approaches will play an essential role for amalgamating these cross-disciplinary interpretations, for redefining acquisition procedures to close existing data gaps, and for optimizing information extraction using image processing methodologies to make the most accurate prognoses of where to find previously undiscovered natural resources. The key task to establishing a foundation for multivariate data-driven analyses relies on developing integration concepts tailored to the available data on various spatial scales and linking these within established data science workflows. The challenge is to acquire all the needed data in a spatially correct context and apply ML on this small training data set.

Here, we present an example from seafloor massive sulphide (SMS) detection at the Trans-Atlantic Geotraverse (TAG) hydrothermal field. Various sources of marine data including autonomous underwater vehicle (AUV) bathymetry and magnetics (Petersen, 2019), and seafloor conductance data derived from Controlled-Source Electromagnetic (CSEM) inversion models are used (Gehrmann et al., 2019).

SMS indicators include a distinct bathymetric manifestation, magnetic low, and high electrical conductance. The latter is likely most indicative of mineral accumulations on the seafloor but only exists along 2D profiles crossing the measurement area due to the logistical expense of acquiring such data. As a result, robust extrapolation of sparsely sampled conductance data onto a regional scale seems efficient for predicting further occurrences of SMS by integrating the acquired bathymetric and magnetic data into a data science framework. This can help improve current predictions of available SMS on the seafloor and provide high-priority site predictions for future validation and sampling campaigns.

The available data allows us to use both unsupervised and supervised machine learning strategies to (a) classify the seafloor based on the spatially distributed bathymetry and magnetic data helping to identify regions with similar characteristics as the known SMS sites; (b) extrapolate conductance data to predict possible SMS sites outside of the 2D CSEM profile lines. Additionally, the high-resolution bathymetry data allows us to enhance our spatial feature matrix through image processing techniques, i.e. edge detection, circular Hough-transforms, and Gabor filtering to improve our spatial understanding of bathymetric features and feed these into a sequential application of fuzzy clustering with random forest regression. The spatially sampled maps are used to create a segmented map of the seafloor combining regions of similar behaviour into common clusters. The pixel fuzziness, which describes the affiliation of each pixel to the corresponding cluster, is then used in a random forest regression approach at the defined locations of the sparsely sampled conductivity data to derive a model that allows us to extrapolate the sparsely sampled conductance data onto a regional scale (Fig. 6.12). Such two-step strategy deprives the ML kernel any physical meaning and relates its predictions solely to the learned patterns.

The results of this pilot study show that unsupervised and supervised machine learning strategies can be used to not only classify the seafloor into regions with similar behaviour, but also identify and predict known and unknown SMS sites in almost real-time. Thus, machine learning provides a robust framework to integrate multivariate data based solely on data-driven analyses, which will be of benefit to marine sciences to (a) optimize marine sampling campaigns through targeted point-scale measurements at regions of greatest interest defined through spatially

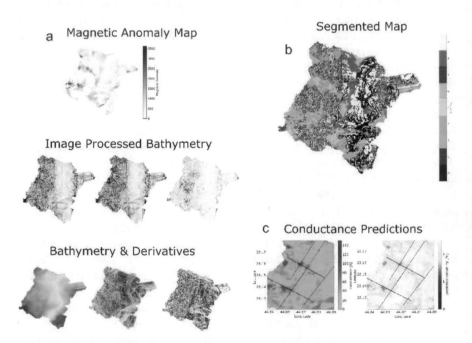

Fig. 6.12 Schematic of ML workflow for predicting SMS sites using multivariate spatial data. (a) Enhanced input features for unsupervised fuzzy clustering, consisting of regional bathymetry data and its derivatives (e.g. slope, ruggedness, aspect), feature enhanced image processed bathymetry (i.e. edge detection, circular Hough Transform and Gabor filtering), and physical property data (magnetic anomaly map). (b) Segmented output map imaging the main contributing component of each pixel. (c) Extrapolated CSEM conductance derived from random forest regression and the corresponding prediction variability assuming 5% Gaussian error. The black markers denote the actual profiles of the CSEM conductance data

distributed geophysical, geochemical, geomorphological, oceanographic or geological data; (b) update first-order predictions of available strategic metals on the seafloor through guided geophysical interpretations (Galley et al. 2021).

6.3.4.3 Deep Neural Networks for Total Organic Carbon Prediction and Data-Driven Sampling

The oceans comprise about 72% of the earth surface and due to its size and available technology, direct seafloor samples collected so far are sparse in space. The existing data sets on sediment composition are inadequate to quantify the fluxes of carbon and other seawater constituents across the seabed globally. Sediment and ocean models are strongly relying on these fluxes to simulate the uptake of atmospheric CO_2 and the biogeochemical cycles in the ocean. Moreover, sampling campaigns are often restricted by ship time, funds, and the lack of consistent methodologies to collect and process the data. Thus, the challenge is to find methods that allow to predict the total organic carbon (TOC) content everywhere in the ocean and show the uncertainty of this prediction with it.

To approach this problem, machine learning methods were adapted to marine sciences to approximate the seafloor physical and biogeochemical properties without the need of direct sampling. Some of these methods (e.g. k-Nearest Neighbours) provide a sophisticated averaging tool to estimate the seafloor property based on the data points nearest in space. However, this approach performs better in more homogeneous environments, which does not apply to global-scale problems.

Over the past decade, deep learning has been used to solve various regression and classification tasks (LeCun et al., 2015). Compared to classical machine learning approaches (k-Nearest Neighbours, Random Forests, etc.), deep learning algorithms excel at learning complex, non-linear internal representations in part due to the highly over-parameterized nature of their underlying models. This advantage often comes at the cost of interpretability. Exemplarily we used deep neural networks (DNN) to assess the TOC content of the global seafloor surface (Fig. 6.13). Implementing Softmax distributions on implicitly continuous data (regression tasks), we obtain probability distributions which can be used to quantify the model's intrinsic information. A variation of the Dropout method, i.e. the Monte Carlo Dropout, is used during the inference step providing a tool to model prediction reliability. Using transfer learning techniques, the resulting model was modified to also make sedimentation rate predictions; sedimentation rate ultimately relates to the problem related of calculating seafloor TOC.

We used these techniques to create model information maps that are a key element in developing new data-driven sampling strategies for data acquisition. Mapping prediction probabilities provide a quantitative analysis of the model information and allows us to define geographical locations that are under-sampled. By acquiring new information at these selected coordinates during upcoming research cruises potentially as part of new global sampling programmes will overall strongly and quickly improve global predictions.

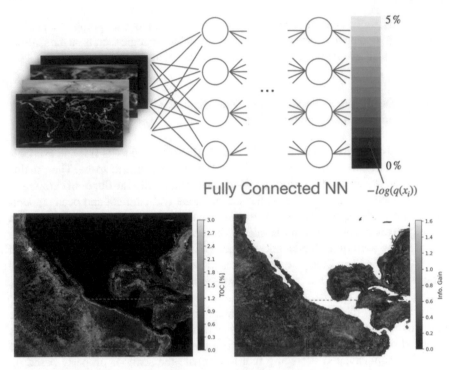

Fig. 6.13 Top: A fully connected neural network with a Softmax activation layer outputs a probability distribution of which the maximum corresponds to the predicted regression value. Bottom: Each prediction point is accompanied by an expected information gain sampling. Often times unexpected/counterintuitive prediction values are tied to a higher expected information gain values (here: low TOC patch in the Pacific Coast of Central America). This points to a higher model uncertainty for the region

Using the prediction probabilities to calculate the information gain from sampling, we were able to generate global maps that can aid data-driven sampling in the future. These information gain maps might be used by scientist to derive the most beneficial decisions on next sampling locations and potentially supports scientific research vessels of opportunity to collect data "during transit" when they pass one these important locations.

References

Alharbi N, Soh B (2019) Roles and Challenges of Network Sensors in Smart Cities, 2019 International Conference on Smart Power & Internet Energy Systems, IOP Conf. Series: Earth and Environmental Science322 (2019) 012002, https://doi.org/10.1088/1755-1315/322/1/012002.

ASQ (2021) Quality Assurance & Quality control, retrieved from https://asq.org/quality-resources/quality-assurance-vs-control on February 19, 2021.

Bjørnstad ON, Shea K, Krzywinski M et al (2020) Nat Methods 17:455–456. https://doi.org/10. 1038/s41592-020-0822-z

Bruns T, Eichstädt S (2018) A smart sensor concept for traceable dynamic measurements. J Phys: Conf Ser 1065: 212011. https://doi.org/10.1088/1742-6596/1065/21/212011

CEN (2021) CEN Workshop Agreement , retrieved from https://boss.cen.eu/developingdelivera bles/CWA/Pages/ on February 19, 2021.

Dehning J, Zierenberg J, Spitzner F P, Wibral M, Neto J P, Wilczek M, Priesemann V (2020) Inferring change points in the spread of COVID-19 reveals the effectiveness of interventions, Science, 369: eabb9789.

Devaraju N, Bala G, Modak A (2015) Effects of large-scale deforestation on precipitation in the monsoon regions: Remote versus local effects. Proc Natl Acad Sci USA 112:3257–3262 https:// doi.org/10.1073/pnas.1423439112

Dietrich P, Leven C (2006) Direct Push Technologies. In: Kirsch R (ed) Groundwater Geophysics. Springer, pp 321–340

Dong E, Du H, Gardner L (2020) An interactive web-based dashboard to track COVID-19 in real time. Lancet Inf Dis. 20(5):533–534. https://doi.org/10.1016/S1473-3099(20)30120-1

Dorigo W, Himmelbauer I, Aberer D, Schremmer L, Petrakovic I, Zappa L, Preimesberger W, Xaver A, Annor F, Ardö J, Baldocchi D, Blöschl G, Bogena H, Brocca L Calvet JC, Camarero J J, Capello G, Choi M, Cosh M C, Demarty J, van de Giesen N, Hajdu I, Jensen KH, Kanniah KD, de Kat I, Kirchengast G, Rai PK, Kyrouac J, Larson K, Liu S, Loew A, Moghaddam M, Martínez Fernández J, Mattar Bader C, Morbidelli R, Musial JP, Osenga E, Palecki MA, Pfeil I, Powers J, Ikonen J, Robock A, Rüdiger C, Rummel U, Strobel M, Su Z, Sullivan R, Tagesson T, Vreugdenhil M, Walker J, Wigneron JP, Woods M, Yang K, Zhang X, Zreda M, Dietrich S, Gruber A, van Oevelen P, Wagner W, Scipal K, Drusch M, and Sabia R The International Soil Moisture Network: serving Earth system science for over a decade, Hydrol. Earth Syst. Sci. Discuss. [preprint], https://doi.org/10.5194/hess-2021-2, accepted, 2021.

Duval E (2001) Metadata Standards: What, Who & Why. J. UCS. 7:591–601

Fischer P (2020) "Intelligent Sensor Technology: A 'Must-Have' for Next-Century Marine Science," In AI Technology for Underwater Robots, (eds.), KF, SS, KD & HN: Springer, 19–36. https:// doi.org/10.1007/978-3-030-30683-0_2

Galley C, Lelièvre P, Haroon A, Graber S, Jamieson J, Szitkar F, et al. (2021) Magnetic and Gravity Surface Geometry Inverse Modelling of the TAG Active Mound. J Geophys Res Solid Earth, 126, e2021JB022228 https://doi.org/10.1029/2021JB022228

Gehrmann R, North LJ, Graber S, Szitkar F, Petersen S, Minshull TA, Murton BJ (2019) Marine mineral exploration with controlled source electromagnetics at the TAG hydrothermal field, 26°N Mid-Atlantic ridge. Geophys Res Lett 46:5808–5816 https://doi.org/10.1029/2019GL082928

Gerchow P, Koppe R, Macario A, Haas A, Schäfer-Neth C. and Pfeiffenberger H. (2015): O2A: A Generic Framework for Enabling the Flow of Sensor Observations to Archives and Publications, European Geosciences Union, Vienna, 12 April 2015—17 April 2015

Gower JC (1971) A General Coefficient of Similarity and Some of Its Properties. Biometrics 27:857– 871 https://doi.org/10.2307/2528823

Gower JC, Legendre P (1986) Metric and Euclidean properties of dissimilarity coefficients. J Classif 3:5–48 https://doi.org/10.1007/BF01896809

Green S (2003) Metadata: Essential Standards for Management of Digital Libraries, ALI Digital Library Workshop Linda Cantara, Metadata Librarian Indiana University, Bloomington, downloaded from https://slideplayer.com/slide/7641133/ on November 12, 2020

Higgins S (2007) DCC Standards Watch 1: What are Metadata Standards?, downloaded from https:// www.dcc.ac.uk/guidance/briefing-papers/standards-watch-papers/using-metadata-standards on November 16, 2020

Höppner F, Klawonn F, Kruse R, Runkler T (2000) Fuzzy Cluster Analysis: Methods for Classification, Data Analysis and Image Recognition. J Oper Res Soc 51 https://doi.org/10.2307/254022

Huffman GD, Bolvin D, Braithwaite K, Hsu R, Joyce Xie P (2014) Integrated Multi-satellitE Retrievals for GPM (IMERG), version 4.4. NASA's Precipitation Processing Center

Kiran M, Murphy P, Monga I, Dugan J (2015) and Baveja SS "Lambda architecture for cost-effective batch and speed big data processing." IEEE International Conference on Big Data (big Data) 2015:2785–2792 https://doi.org/10.1109/BigData.2015.7364082

Knapp KR, Diamond HJ, Kossin JP, Kruk MC, Schreck CJ (2018) International Best Track Archive for Climate Stewardship (IBTrACS) Project, version 4. NOAA National Centers for Environmental Information https://doi.org/10.25921/82ty-9e16

Knapp KR, Kruk MC, Levinson DH, Diamond HJ, Neumann CJ (2010) The International Best Track Archive for Climate Stewardship (IBTrACS): Unifying tropical cyclone bst track data. Bull Am Meteor Soc 91:363–376 https://doi.org/10.1175/2009BAMS2755.1

Koedel U, Schuetze C, Fischer FP, Bussmann I, Sauer PK, Nixdorf E, Kalbacher T, Wiechert V, Rechid D, Bouwer LM, Dietrich P (2022) Challenges in the evaluation of observational data trustworthiness from a data producers viewpoint (FAIR+). Front Environ Sci 9:art. 772666 https://doi.org/10.3389/fenvs.2021.772666

Koppe R, Gerchow P, Macario A, Haas A, Schäfer-Neth C, Pfeiffenberger H (2015) O2A: A generic framework for enabling the flow of sensor observations to archives and publications, OCEANS 2015 - Genova. Genova, Italy 2015:1–6 https://doi.org/10.1109/OCEANS-Genova. 2015.7271657

Kunkel KE, Karl TR, Easterling DR, Redmond K, Young J, Yin X, Hennon P (2013) Probable maximum precipitation and climate change. Geophys Res Lett 40:1402–1408 https://doi.org/10. 1002/grl.50334

LeCun Y, Bengio Y, Hinton G (2015) Deep learning. Nature 521:436–444 https://doi.org/10.1038/ nature14539

Lombard et al. (2019). Globally Consistent Quantitative Observations of Planktonic Ecosystems. Front. Mar. Sci., https://doi.org/10.3389/fmars.2019.00196

Olteanu A, Vieweg S, Castillo C, 2015 February: What to expect when the unexpected happens: Social media communications across crises. In Proceedings of the 18th ACM conference on computer supported cooperative work & social computing, pp. 994–1009. Association for Computing Machinery https://doi.org/10.1145/2675133.2675242

Paasche H, Paasche K, Dietrich P (2020) Nature and Culture 15:1–18 https://doi.org/10.3167/nc. 2020.150101

Paasche H, Tronicke J (2007) Cooperative inversion of 2D geophysical data sets: A zonal approach based on fuzzy c-means cluster analysis. Geophysics 72:A35–A39 https://doi.org/10.1190/1.267 0341

Paasche H, Tronicke J, Dietrich P (2010) Automated integration of partially colocated models: subsurface zonation using a modified fuzzy c-means cluster algorithm. Geophysics 75:P11–P22 https://doi.org/10.1190/1.3374411

Paasche H, Tronicke J, Holliger K, Green AG, Maurer H (2006) Integration of diverse physical-property models: Subsurface zonation and petrophysical parameter estimation based on fuzzy c-means cluster analyses. Geophysics 71:H33–H44

Petersen S, (2019): Bathymetric data products from AUV dives during METEOR cruise M127 (TAG Hydrothermal Field, Atlantic). GEOMAR - Helmholtz Centre for Ocean Research Kiel, PANGAEA, https://doi.org/10.1594/PANGAEA.899415

Sauerer J, (2013) Smart Sensors, Sensorik für erneuerbare Energien und Energieeffizienz : Beiträge zum Workshop vom AMA Fachverband für Sensorik e.V. und vom ForschungsVerbund Erneuerbare Energien am 12. und 13. März 2013 in Berlin-Adlershof Berlin, 2013, https://www.fvee.de/fil eadmin/publikationen/Workshopbaende/ws2013/ws2013_03_02.pdf, downloaded on November 14, 2020

Schmidhalter U, Maidl F-X, Heuwinkel H, Demmel M, Auernhammer H, Noack P, Rothmund M (2008) Precision Farming—Adaptati on of land use management to small scale heterogeneity. In: Schröder P, Pfadenhauer J, Munch JC (eds) Perspectives for agroecosystem management, Elsevier, pp 121–199

Schrön M, (2017) Cosmic-ray neutron sensing and its applications to soil and land surface hydrology (PhD thesis). Potsdam, Germany: University of Potsdam

Schrön M, Zacharias S, Womack G, Köhli M, Desilets D, Oswald SE, Bumberger J, Mollenhauer H, Kögler S, Remmler P, Kasner M, Denk A, Dietrich P (2018) Intercomparison of cosmic-ray neutron sensors and water balance monitoring in an urban environment. Geosci. Instrum. Method. Data Syst. 7(1):83–99

Schultz MG, Kunkel R, Petzold A, (2019). New perspectives on quality assurance and quality control of environmental observation data, D.E. Newsletter December 2019, downloaded on https://www.digitalearth-hgf.de/storage/379/Newsletter_DE_2019_12_FZJ.pdf November 28, 2020

Schwarze R, Herrmann A, Münch A ,Grünewald U, Schöniger M (1991) Rechnergestützte Analyse von Abflusskomponenten und Verweilzeiten in kleinen Einzugsgebieten. 35:143-184

Spencer B, Ruiz Sandoval M, Kurata N (2004) Smart Sensing Technology: Opportunities and Challenges. Struct Control Health Monit 11:349–368 https://doi.org/10.1002/stc.48

Steinacker A, Ghavam A, Steinmetzt R, (2001) Metadata Standards for web-based Resources. IEEE Multimedia, downloaded at http://ivizlab.sfu.ca/arya/Papers/IEEE/Multimedia/2001/Jan/Metadata%20Standards%20for%20Web-based%20Resources.pdf on November, 17, 2020

Thakur D, Kumar Y, Kumar A, Singh P (2019) Applicability of Wireless Sensor Networks in Precision Agriculture: A Review. Wireless Pers Commun 107 https://doi.org/10.1007/s11277-019-06285-2

UFZ (2021). Modular Observation Solutions for Earth Systems (MOSES), retrieved from https://www.ufz.de/moses/ on February 19, 2021

Ullo SL, Sinha GR (2020) Advances in Smart Environment Monitoring Systems Using IoT and Sensors. Sensors (basel, Switzerland) 20(11):3113 https://doi.org/10.3390/s20113113

Vicente-Serrano SM, Beguería S, López-Moreno JI, (2010) A Multi-scalar drought index sensitive to global warming: The Standardized Precipitation Evapotranspiration Index - SPEI. J Clim , 23 (7):1696–1718 https://doi.org/10.1175/2009JCLI2909.1

Viscarra Rossel RA, Adamchuk VI, Sudduth KA, McKenzie NJ, Lobsey C (2011) Proximal soil sensing: An effective approach for soil measurements in space and time. In: Donald L. Sparks, (ed) Advances in Agronomy, Vol. 113, Burlington: Academic Press, pp 237–282. ISBN: 978-0-12-386473-4, Elsevier Inc. Academic Press

Wang Y, Wang T, Ye X, Zhu J, Lee J (2016) Using social media for emergency response and urban sustainability: A case study of the 2012 Beijing rainstorm. Sustainability 8(1):25 https://doi.org/10.3390/su8010025

Wilkinson M, Dumontier M, Aalbersberg I et al (2016) Sci Data 3:160018 https://doi.org/10.1038/sdata.2016.18

Zacharias S, Bogena H, Samaniego L, Mauder M, Fuß R, Pütz T, Frenzel M, Schwank M, Baessler C, Butterbach-Bahl K, Bens O, Borg E, Brauer A, Dietrich P, Hajnsek I, Helle G, Kiese R, Kunstmann H, Klotz S, Munch JC, Papen H, Priesack E, Schmid HP, Steinbrecher R, Rosenbaum U, Teutsch G, Vereecken H (2011) A network of terrestrial environmental observatories in Germany. Vadose Zone Journal 10(3):955–973 https://doi.org/10.2136/vzj2010.0139

Zhang D, Eng B, Prof S, Connor NEO, Regan PF, Ph.D. Thesis. Dublin City University; Dublin, Ireland: 2015. Multi-Modal Smart Sensing Network for School of Electronic Engineering

Zhang AB, Gourley D, (2009) Creating metadata, In Chandos Information Professional Series

Zreda M, Shuttleworth WJ, Zeng X, Zweck C, Desilets D, Franz T, Rosolem R (2012) COSMOS: The COsmic-ray soil moisture observing system. Hydrol Earth Syst Sci 16:4079–4099 https://doi.org/10.5194/hess-16-4079-2012

Chapter 7
Interdisciplinary Collaboration

Nike Fuchs and Gesche Krause

Abstract The Digital Earth Project aims at a strong interdisciplinary collaboration of the various Earth science disciplines and data science, to foster digitalization and the application of data science methods. As this is a highly complex interdisciplinary endeavour that involves eight research centres and many scientists, a success evaluation was deployed after the first half of the project. A social science-oriented evaluation was conducted, in which a World Cafe and a survey were used to evaluate the success of the collaboration and opportunities for improvement. Results indicate a strong need among participating scientists to more clearly understand and advocate for the overarching goals, have more face-to-face interaction, optimize the use of existing research infrastructure, and develop a sound perspective for knowledge transfer and long-term continuation of the developed approaches. It was deduced that individuals shape the process and that digitization is more than just a technical matter, but depends heavily on individuals and the process of implementation.

Keywords Evaluation · World Cafe · Survey · Collaboration · Interdisciplinary · Earth System Science

7.1 Challenges

For Digital Earth, one of the biggest challenges was bridging the gap between different disciplines and achieving the project goals in an extremely heterogeneous environment of project partners, scientific concepts and vocabularies. The consortium decided to seek support from the authors as representatives of the social sciences who are scientifically concerned with interactions in heterogeneous groups and to examine and assess the interdisciplinary collaboration. This chapter presents the results of a World Café conducted with Digital Earth scientists from a social science

N. Fuchs (✉) · G. Krause
Alfred Wegener Institute Helmholtz Centre for Polar and Marine Research, Bremerhaven, Germany
e-mail: nike.fuchs@awi.de

© The Author(s) 2022
L. M. Bouwer et al. (eds.), *Integrating Data Science and Earth Science*,
SpringerBriefs in Earth System Sciences,
https://doi.org/10.1007/978-3-030-99546-1_7

perspective to learn more about pitfalls, challenges, requirements and best practices for successful collaboration.

Mankind on the threshold of the digital age is facing fundamental challenges in the expansion of opportunities and further development of even more far-reaching key technologies. The core characteristics of the digital age, namely networking, cognition, autonomy, virtuality and the explosion of knowledge (Schieferdecker and Messner 2019) have embraced the scientific world long since.

Over the recent decades, emphasis has been placed on making the scientific process more open and inclusive for all relevant actors, within and beyond the scientific community, as enabled by digitalization (Dai et al. 2018). That said, digitalization is changing science fundamentally. This poses the challenge that different scientific communities have developed their own vocabulary, observation methods, concepts and models that need to be brought together to advance on the required digitalization and integration.

This growing plurality of knowledge can be also observed in the realm of Earth system science, in which—the current research has branched off in multiple detailed sub-disciplines that call for new forms of collaboration across the different research strands. In this context, digitalization is believed to play a central role in this effort to tie the "loose" ends. Undertaking digitalization within Earth system science, however, involves large amounts of data, necessitating streamlining across different scientific communities, which can offer new analytical possibilities and produces new sorts of decision support tools. The moment an innovation process such as digitalization is initiated, the organization on which it is brought onto undergoes an initial phase, which may appear chaotic. This stage, dubbed as "fuzzy front-end of innovation", plays a decisive role in the further roll out of this innovation process (Berghaus and Back 2017).

The speed of uptake of digitalization is determined by the way how the (science) network community deals with the new demands (Clegg et al. 2016). As a case in point, with starting the Digital Earth project, an already existing scientific community was challenged with a completely new situation; to conduct and advance "data science" with a set with unclear parameters. In general, such challenges entail the adaptation and alteration of user behaviour (Brenner et al. 2014), and the accessibility and usability of data and newly introduced technologies (Dery and MacCormick 2012; Berghaus and Back 2017). On a social level, during the institutionalizing of innovation, new practices, values, routines and social norms have to be developed; networks are powerful carriers for this (Clegg et al. 2016).

While most challenges of the Digital Earth project were clear from the onset, others surfaced through interaction with others and through collaborative reflection. A validation on a personal level is thus required to link system perspectives and worldviews with research approaches and to assess efficacy of collaboration (Chiocchio et al. 2012; Glassman et al. 2021). Indeed, engaging with other fields of research can be a time-consuming process. To facilitate the gap-bridging of the different knowledge realms, one tool is the world cafe method (Brown et al. 2010). It provides the opportunity to jointly identify the challenges and gain shared consensus together as a group. Furthermore, this consensus and related challenges are not only shared and validated, but also recorded and by that formally acknowledged among the group as a whole.

The objective of this chapter is to present the results of an accompanying research evaluation, focussing on the social dimensions on the collective and individual level of the challenges in collaboration within the Digital Earth project. The results have shown several issues that can be improved on and to help address several of the above mentioned challenges.

7.2 Material and Methods

The present research was conducted by using a mixed-methods approach (Kelle 2014). An earlier survey, performed in April 2019 by the Digital Earth project for success evaluation, addressed the collaboration success by the then present requirements for data science, the scientific and project successes and the usability of results at that time (see also Chapter 6 in this book). The quality of collaboration was not assessed at that time. To examine the success of collaboration within Digital Earth in more detail, an online survey, more focussed on the social dynamic across different scientific disciplines, was conducted prior the 2nd Interim Meeting of the project in January 2020, as the was half-way. In this second survey, qualitative and quantitative metrics were deployed to identify potential collaboration barriers after Hanson (Hanson 2009). The findings of the online poll formed the pre-assessment stage which acted as baseline for the successive assessment steps. As such, during the 2nd Interim Meeting of the project in January 2020 itself, a World Cafe was conducted to assess trends and nature of collaboration among all attendees in more depth. The World Café, and to come up with proposals for potential improvements that would lead to better collaboration is a large group method, which contains a sequence of discussions at tables with 4–7 people seated at each table (Brown et al. 2010). The Digital Earth World Café consisted of 3 rounds, with each 3 questions, two of them in two versions, thus 5 questions in total. 49 scientists, engaged in Digital Earth, devoted effort in addressing those questions during the World Cafe. The questions evolved around the approaches and tools of collaboration, trajectories and trends, as well as on potential next steps. Central focus of the exercise was to gain insights on individual and collective views on the collaboration, and thus success, within the project. Also, potential areas for improvement for collaboration were identified.

7.3 Results and Discussion

In the following, highlights of the World Café discourses are collated and presented in a summative manner. **In round 1 and 2**, respectively, focus was placed on approaches and tools of collaboration as well as emerging trajectories and trends of collaboration.

It is a noteworthy aspect, that it was possible to distil four major thematic aspects across the first two rounds from the collected statement pieces. These 4 major groups were confirmed and strengthened during the final prioritization round:

1. *Project Goals*: a frequent mentioning and a clear feedback in the voting session suggested that not all participants were able to see the higher level and overall goals that were set for the project and hence voiced a wish for a clearer definition.
2. *Individual Level*: the wish for more personal interaction, interpersonal exchange and cohesion was clearly voiced and appeared in the statements on all tables and resurfaced in each World Café round.
3. *Infrastructure*: Although available access, clarity of structure within and understanding of the used platforms was identified as a major component of good collaboration, the use of the digital infrastructure (closer defined as GITHUB, Helmholtz Net, Confluence) and therefore the exchange of information between Centres was mentioned as a major barrier for collaboration. The major obstacles here appear to be the optimal use of the infrastructure which has been developed for the project. This includes the lack of sufficient overview of the various platforms, search options for people and information, guidelines for use and communication about these infrastructures.
4. *Knowledge Transfer and Continuation*: the participants expressed repeatedly the wish to see the application of the already produced project outcomes as well as the outreach to increase visibility for their product, and furthermore the continuation of the project after the prospective project end.

In **round 3,** a special focus was placed on the next steps in collaboration within Digital Earth, the results given to the question "In terms of collaboration, what should we do next?" showed the same prevailing dispositioning as in the first two rounds and were sorted in strategic and methodological suggestions or advice (Table 7.1).

Some of the statements are at the interface between two dimensions, e.g. "Introduce new members to everyone", applies to the infrastructural dimension, helping to find the right contact person, but also feeds into the personal dimension. "Develop and implement long-term-legacy plans" touches not only outreach and the big picture, but also the goals. The previously conducted online poll, aiming at the identification of potential collaboration barriers, mirrored this finding, as it showed that half of project members indicated that within a geographically wide-spread team, finding the right contact person is difficult. Furthermore, it supported the finding of an abundantly stated request for a clearer and more personal level networking as well as a high commitment and willingness to collaborate. This was somewhat reflected in the World Café's last round, in which ideas about next steps for future collaboration were collected and needed little support by the facilitator, as the notions seemed to be clear to the participants and motivation was high. In conclusion, the participant's statements during the World Café showed an overall coherence in their professional needs and the challenges which the project faces in the 2nd half. Intensified personal contact and subsequently refined alignments towards the mentioned themes were identified as prerequisites for the project's success. Furthermore, the wish for clearer defined goals and targets were highly abundant throughout the entire World Cafe. The

Table 7.1 Sorted statements from World Café round 3 in response to "In terms of collaboration, what should we do next?"

Dimension	Level	Statements given by participants
Individual	Strategic	– Communication of results, workflows, methods, mistakes, lessons learned – Don't settle with current status of collaboration, learn more ways to work with and familiarize yourself with others – Consolidate a shared language and common understanding
	Methodological	– Create a forum or board for exchange – Create examples for successful collaboration – Meeting more often, create opportunities to exchange, also among disciplines, because we learn better from others when meeting in person – Create an alumni network with regular meetings (e.g. DigitalEarth@Kieler Week)
Goals	Strategic	– We need common goals and targets – Define clear milestones – Reiterate common goals and communicate them
	Methodological	– Regular meetings with others to realign towards goals – Consolidation of targets and activities towards common goals (steering tasks)
Infrastructure	Strategic	– Make information available to participants – One selling point for information
	Methodological	– Promote existing platforms better, especially among new people – Integrate tools towards one framework – Overcome technical barriers (AAI, exchange of resources) – Component-based software implementation and design (defined interfaces) – Finding partners/people with similar issues search engine, look-up table, Digital-Earth-Wiki – Joint surveys and collaborative inter-comparison campaigns – Make use of the agreed methods and standards for communication, collaboration and co-working – Guidelines for using collaboration system – Advertise information sources to participants more often – Introduction of new members to everyone

(continued)

Table 7.1 (continued)

Dimension	Level	Statements given by participants
Outreach and Knowledge Transfer	Strategic	– Develop and implement long-term legacy plans – Reiterate the connections in the big picture – Find ways to continue collaboration post-Digital-Earth – Increase active engagements in larger scale data science/management activities – What do other projects need and what can we offer?
	Methodological	– Engage with other initiatives: what can Digital Earth do for you? – Define success stories and apply feedback to improve collaboration – Communicate and show results on simple and well-known platform (YouTube) – Document more: publicly and discoverable – Encourage Digital-Earth-members to broadcast their recent To-do's, tasks, aims, issues and problems – More visibility for MOSES and others: Campaign and data management – visibility of Digital Earth contributions – Who's who? (like google street view)

way how the shared web spaces are organized and the overall information accessibility was a common theme as well. The fourth thread was the wish for more outreach activities, the requirement to understand the future application and usage of the products from the project, as well as a clear perspective for the period after the direct project life-time including a possible further development of ideas and products.

7.4 Conclusions and Outlook

The discourses and reflection within the Digital Earth project function as a case study for the conditions in which human societies at large find themselves today; digital participation and networking enable a manifold of potential but also need to adhere to essential social mechanisms. These divergences also surface within science and on the perspectives of how to collaborate and streamline different data towards open science outcomes. The deployment of digital tools and methods alone does not guarantee a successful digitalization.

Two central issues could be identified:

1. Individuals shape the process: In essence, the findings uncover the hidden assumptions and biases each of the individual partaking scientists had regarding digitalization in Earth System Sciences. The discussions in the World Café exercise indicated that the background, experiences and personal knowledge of each individual seem to determine the definitions and views on how to collaborate in the project. Yet, tools may help to streamline some of the diverging initial definitions and ideas expressed at the World Café. In this context, the World Café proved to be a suitable method of positive engagement across different disciplines.

2. Digitalization is more than solely a technical affair and relies heavily on individuals, their understanding of collaboration and a harmonization of disciplinary perspectives and worldviews. While there was general agreement among researchers that biophysical knowledge remains critical in their work, the need for new digital capabilities and clear objectives on how to continue to merge science towards digitalization. The findings indicate that at a high abstraction level, the expectations of digitalization within the project were quite unequivocal across the different research disciplines. For instance, a similar understanding was portended that digitalization potentially leads to more productive, efficient and sustainable forms of data utilization and knowledge creation. However, this understanding of digitalization was hampered by the formulated need for a clearer definition of the related and required digitalization process within the different research organizations, which suggests that the project was in somewhat earlier stages of "digi-grasping" (Dufva and Dufva 2019) or what has been referred to as the "fuzzy front end of digitalization" (Berghaus and Back 2017).

In the light of the global challenges ahead, combined with the possibilities and requirements of the dawn of the digital age, not enough emphasis can be laid on the investment in the underlying personal connections. The foundation of all interactions, always had and most likely will be for a long time, is the connection on an individual, personal level. By acknowledging this, digitalization in Earth System Science can, and most likely will be a highly potential tool for fostering meaning-making and understanding of the complex world around us.

Acknowledgements The authors would like to express their gratitude to all members of the Digital Earth project for their willingness to engage and share their views.

References

Berghaus S, Back A (2017) Disentangling the fuzzy front end of digital transformation: Activities and approaches. In: ICIS 2017 Proceedings, AIS Electronic Library, 2017.

Brenner W, Karagiannis D, Kolbe L et al (2014) User, use & utility research. Bus & Inf Syst Eng 6:65–72. https://doi.org/10.1007/s12599-013-0302-4

Brown J, Isaacs D, World Café Community (2010) The world café: shaping our futures through conversations that matter. Berrett-Koehler Publishers, San Francisco

Chiocchio F, Grenier S, O'Neill T et al (2012) The effects of collaboration on performance: a multilevel validation in project teams. Int J Proj Organ Manag 4(1):1–37. https://doi.org/10.1504/IJPOM.2012.045362

Clegg S, Josserand E, Mehra A et al (2016) The transformative power of network dynamics: a research agenda. Organ Stud 37(3):277–291. https://doi.org/10.1177/0170840616629047

Dai Q, Shin E, Smith C (2018) Open and inclusive collaboration in science: a framework. OECD Science, Technology and Industry Working Papers 2018/07. https://doi.org/10.1787/2dbff737-en

Dery K, MacCormick J (2012) Managing mobile technology: the shift from mobility to connectivity. MIS Q Exec 11(4):159–173

Dufva T, Dufva M (2019) Grasping the future of the digital society. Futures 107:17–28. https://doi.org/10.1016/j.futures.2018.11.001

Glassman M, Kuznetcova I, Peri J et al (2021) Cohesion, collaboration and the struggle of creating online learning communities: Development and validation of an online collective efficacy scale. Comput Educ Open 2. https://doi.org/10.1016/J.CAEO.2021.100031

Kelle U (2014) Mixed methods. In: Baur N, Blasius J (ed) Handbuch Methoden der empirischen Sozialforschung. Springer VS, Wiesbaden, pp 153–166. https://doi.org/10.1007/978-3-531-18939-0_8

Schieferdecker I, Messner D (2019) Die digitale Nachhaltigkeitsgesellschaft. https://deutschland-und-die-welt-2030.de/de/beitrag/dei-digitalisierte-nachhaltigkeitsgesellschaft/. Accessed 20 October 2021

Hansen MT (2009) Collaboration: how leaders avoid the traps, create unity, and reap big results. Harvard Business Press, Cambridge, MA

Rijswijk K, Klerkx L, Turner JA (2019) Digitalisation in the New Zealand Agricultural Knowledge and Innovation System: initial understandings and emerging organisational responses to digital agriculture. NJAS-Wagening J Life Sci 90:100313

Chapter 8
Evaluating the Success of the Digital Earth Project

Laurens M. Bouwer, Diana Rechid, Bernadette Fritzsch, Daniela Henkel, Thomas Kalbacher, Werner Köckeritz, and Roland Ruhnke

Abstract The Digital Earth project aims at a strong interrelation between Data and Earth Science and a step-change in implementing data science methods within Earth science research. During the project, the progress of interdisciplinary collaboration and adoption of data science methods has been measured and assessed with the goal to trace the success of the project. This chapter provides the set-up of this evaluation and the results from two online questionnaires that were held after the start and before the end of the project.

Keywords Evaluation · Collaboration · Digitalisation · FAIR · Data science · Capacities

8.1 Objective

The Digital Earth project addresses the challenge of digital transformation and adoption of data science methods in Earth sciences. Therefore, its focus is on linking natural science and data science and to develop approaches for (i) data

L. M. Bouwer (✉) · D. Rechid
Climate Service Center Germany (GERICS), Helmholtz-Zentrum Hereon, Hamburg, Germany
e-mail: laurens.bouwer@hereon.de

B. Fritzsch
Alfred Wegener Institute Helmholtz Centre for Polar and Marine Research, Bremerhaven, Germany

D. Henkel
GEOMAR Helmholtz Centre for Ocean Research Kiel, Kiel, Germany

T. Kalbacher
Helmholtz Centre for Environmental Research-UFZ, Leipzig, Germany

W. Köckeritz
Helmholtz Centre Potsdam-GFZ German Research Centre for Geosciences, Potsdam, Germany

R. Ruhnke
Karlsruhe Institute of Technology, Eggenstein-Leopoldshafen, Germany

© The Author(s) 2022
L. M. Bouwer et al. (eds.), *Integrating Data Science and Earth Science*,
SpringerBriefs in Earth System Sciences,
https://doi.org/10.1007/978-3-030-99546-1_8

analysis and exploration; (ii) data collection and monitoring; and (iii) interdisciplinary collaboration, which is of special importance in the digital transformation (see Chap. 2).

During the Digital Earth project, the success of adopting data science methods in the field of Earth sciences has been shown, as well as the scientific progress that can be achieved by doing this. Chapters 3, 4, 5 and 6 of this book give several examples of this. In addition, the collaboration between the different Earth Science fields and the different research centres involved in the project, and specifically, the collaboration between the Earth Sciences and data science disciplines was another important focus. Thus, we wanted to evaluate the process of interdisciplinary research and the application of data science methods and how this has evolved during the project.

Project monitoring and evaluation are an important process, to identify challenges before and during the project, and to reflect and improve the research project outcomes, learn and adjust activities during the project lifetime, and also to set clear goals for follow-up activities or projects afterwards. Such an evaluation was also deemed useful for the Digital Earth project, as it was the goal to deliver a step-change in the use of data science methods within the different fields of Earth Sciences within the Helmholtz Association. This activity can be seen in the context of the evaluation of other efforts in digitalisation, such as the evaluation of the development of Virtual Research Environments in the United Kingdom (see Junge et al., 2007).

The evaluation aimed at measuring the difference between what the research centres could do at the start of the project, versus what has been achieved after implementation of the Digital Earth project. We also wanted to learn in the course of the project about possible challenges, and the progress and successes of the project, and we wanted to identify possible needs for improvement in process, content and tools during the project.

The approach for this evaluation consisted of the method of online questionnaires. This chapter gives an overview about the online questionnaires; it provides the setting and criteria for evaluation, as well as some of the evaluation results.

8.2 Approach for Evaluation in the Digital Earth Project

The first online questionnaire was done shortly after the start of the project, to establish the important capacities and needs of the research teams to reach success at the start of the project, and to define suitable criteria for measuring such needs. The overarching questions for the evaluation were:

- What are the requirements for data science?
- What is the scientific progress?
- What is the usability of the scientific workflows?
- What is the success of Digital Earth?

These evaluation questions have been used to gather information about the current and desired capacities at the research centres that are involved in the Digital Earth project, as well as about current and anticipated collaboration between the centres, available scientific workflows, digital data and tools and applications. In addition, an assessment was made of the application of the FAIR principles and other measures to enable transfer of results, data and information to other users. The building blocks of the evaluation have included:

1. The development of criteria and indicators to be applied;
2. Development of survey questions;
3. Implementation of a questionnaire at the start of the project;
4. The analysis of the questionnaire results, including an analysis of the current scientific and data standards applied at the centres;
5. Refinement of the criteria and evaluation questions;
6. The repeated monitoring of the progress through a second questionnaire;
7. Reporting in a final assessment report.

Afterwards the questionnaire was analysed, and questions were refined and extended.

8.3 Evaluation Criteria

To develop the questionnaire, we adopted criteria for measuring the project status at the beginning of the Digital Earth project. These overall criteria are listed in Table 8.1 and consist of several sub-criteria. For these sub-criteria, evaluation questions have been formulated for the questionnaire (see Appendix) that have been answered by the members of the Digital Earth consortium from all involved research centres. The main categories are:

- Capacities for doing data science;
- Project success and scientific progress in Digital Earth;
- Usability of the results.

The capacities for doing data science were assessed in order to learn shortly after the start of the project what additional capacities and collaborations may be needed. In addition, we wanted to evaluate how the capacities have improved after the implementation of the project. Therefore, the questions were repeated shortly before the end of the project. The project success and scientific progress were assessed, in order to evaluate expectations and status before and after the project implementation.

Finally, the usability of the results was an important topic. We wanted to assess to what extent the methods, scientific workflows as well as the generated data are usable for the scientific community, but also for other users within society. Here, we used the FAIR principles for scientific data (Wilkinson et al., 2016; Stall et al., 2019) to evaluate how data science implementation is done with the multitude of environmental data that is being used and produced.

Table 8.1 Criteria for assessing success in Digital Earth

Criteria	Capacities for data science	Project success and scientific progress	Usability of results
Sub-criteria	– Capacities – Challenges – Data, infrastructure, models – Data science methods – Data exploration tools – Project collaboration and management	– Scientific goals – Research process goals	– FAIR: Findable, Accessible, Interoperable, Reusable – Usability of results beyond research

FAIR stands for Findable, Accessible, Interoperable and Reproducible. We focus on both observational and model data, as well as the scientific software and tools, as they are the essential basis for the research in the Digital Earth project. The FAIR principles allow for a check on the accessibility, usability and quality of the research results and have become an important basis for scientific practice in all research areas worldwide. The implementation of the FAIR principles for the field of Earth and environmental research has gained importance. Making sure that data are "FAIR" is a major prerequisite for applying data science methods. This is in line with other efforts to advance the FAIRness of digital assets and provide open and seamless access to a set of interoperable FAIR data services, such as through the ENVRI-FAIR project (https://envri.eu/home-envri-fair/). This latter project has developed the "FAIRness" assessment methodology, to evaluate the findability, accessibility, interoperability and reproducibility of provided digital assets, including the datasets that are being provided, but also the scientific methods, workflows and software that have been developed in Digital Earth.

In addition, we have assessed to what extent external users have access to project results and are supported in using the data products, methods, workflows and tools. This includes also users outside academia, thereby underlining the need that Earth and Environmental Science research should also benefit society.

The questionnaire is structured according to the three criteria listed in Table 8.1. In the following sections, we present the three main criteria categories and the related sub-criteria: 1) Capacities for doing data science; 2) Project success and scientific progress; and 3) Usability of results. In the Appendix, the text of the questionnaire is provided with all questions related to the sub-criteria presented below.

8.3.1 Capacities for Doing Data Science

Table 8.2 presents the criteria for measuring the capacities to do data science, and additional requirements for improving those capacities. Each sub-criterion is assessed using a specific indicator. For instance, the sub-criterion "team size" is measured using the indicator "number of persons", which is a quantitative indicator. For other

Table 8.2 Criteria for capacities for doing data science

Sub-criteria	Indicator	Type
Scientific discipline	Scientific discipline types	Quantitative
Sub-discipline	Sub-disciplines	Quantitative
Team size	Number of persons	Quantitative
AI/ML/DL expertise	Number of persons	Quantitative
Advanced Data Visualisation expertise	Number of persons	Quantitative
Data management expertise	Number of persons	Quantitative
Data science capacity in house	Self-assessment	Qualitative
Data science collaboration within the Digital Earth Project	Number of Centres	Qualitative
Data science collaboration externally	Research centres	Qualitative
Data science need for more support	Self-assessment	Qualitative
Collaboration—current	Research centres	Qualitative
Collaboration—additionally required	Research centres	Qualitative
Limitations and challenges	Open question	Qualitative
Observational and model datasets to be used	Open question	Qualitative
Observational infrastructure to be used	Open question	Qualitative
Data and information infrastructure	Open question	Qualitative
Models and data	Open question	Qualitative
New models	Open question	Qualitative
Applied data science methods	Closed question	Qualitative
Data science methods to be applied	Closed question	Qualitative
Purpose for data science methods	Closed question	Qualitative
Data exploration tools	Open question	Qualitative
Requirements for data management	Open question	Qualitative
Best practices for data management	Open question	Qualitative

criteria, qualitative descriptions are used, such as for the description of available Data Science capacities within each group, which is measured along a qualitative scale, ranging from "We are doing fine within our Centre", to "We may need more support from Digital Earth partners or others".

8.3.2 Scientific and Project Goals

Table 8.3 presents the criteria for assessing the scientific and project goals. These criteria are related to the scientific goals that were set in the project and the extent to which the participants in the questionnaires found they were relevant for their work, or have been achieved towards the end of the project. In addition, we analysed the

Table 8.3 Criteria for assessing the scientific and project goals

Sub-criteria	Indicator	Type
Scientific goals	Different scientific goals	Qualitative
Important scientific goals	Ranking of scientific goals	Qualitative
Scientific goals	Key products and tools that will be / have been produced*	Qualitative
Scientific goals	Success in harmonisation and integration of data from different disciplines*	Qualitative
Research process goals	Process goals	Qualitative
Overall success	Planned/delivered joint output (publications, proposals, summer schools, software, data services, etc.)*	Quantitative

*Questions about delivering on goals were only investigated in the final questionnaire

project process goals, as intermediate steps in the research, including quality of data and models, better guidance for field measurements, saving of resources and time, and improved usability of data, information and workflows.

8.3.3 Usability of Results

Table 8.4 presents the criteria for assessing the usability of the project results. They focus specifically on making project results, data and tools available and accessible within and outside the research field. The criteria and indicators are largely based on the FAIR criteria, and related indicators were developed or adopted from the FAIR framework. In particular, we have used several of the metrics developed by Wilkinson et al. (2018).

Table 8.4 Criteria for assessing usability of results

Sub-criteria	Indicator	Type
Findable	Use of DOI, ORCID, IGSN	Quantitative
Accessible	Repositories	Qualitative
Accessible	Metadata description	Quantitative
Accessible	Accessibility of code	Quantitative
Interoperable	Technical data standards	Quantitative
Interoperable	Scripts in formal language	Quantitative
Reusable	Open data and software policies	Quantitative
Reusable	Software and data availability beyond academia	Quantitative
Reusable	Support for data use, including user guidance and user services/advice	Qualitative

8.4 Results from the Questionnaires

The questions listed in Appendix were used to evaluate the status and results of the Digital Earth project through a questionnaire, held after 10 months after the start of the project, and shortly before the project end, after 31 months. The questionnaire was set up in the commercial digital online software package Survey Hero (https://www.surveyhero.com). The questionnaire was filled in without personal details, except for the institution and type of research expertise, and staff role. A total number of 54 respondents of the research staff submitted replies to the questionnaire on both occasions. 47 of these respondents completed most questions in the first questionnaire, and 48 respondents during the second time. A total of 118 invitations were sent in the first questionnaire, which implies a response rate of about 50%, which is reasonable for this type of surveys.

About two-thirds of the research staff that responded identifies themselves as Earth Scientist (Table 8.5). In the second questionnaire, the number of staff identifying as data scientists has increased by half.

The most important results and conclusions from two questionnaires and evaluation within the Digital Earth project include the following:

Several collaborations have been established during the Digital Earth project, during the proposal writing process, and also through collaborations during the project. This is documented through the responses from the researchers in Digital Earth on their collaborations and exchanges with other research centres. In the final questionnaire, each respondent indicated on average 1.7 collaborations with other centres. Towards the end of the project, there was still even more potential and wishes for collaboration reported in the responses. Collaboration with a few specific centres working on data science was highlighted as these are desired for their competences in the field of data science methods.

The project has also progressed on interdisciplinary approaches, showing collaborations between different fields of Earth and Environmental Science, but also tying in data science expertise, in particular related to visual data exploration, Artificial Intelligence, and scientific workflows. In terms of required capacities for doing Data Science, about half of the participants indicate that they have found and benefited

Table 8.5 Scientific disciplines of the questionnaire participants

Scientific discipline	Respondents	
	First questionnaire	Second questionnaire
Earth Scientist*	36	32
Data scientist	15	22
Other**	3	-
Total	**54**	**54**

*Includes biologist and marine biogeochemist
**The category "Other" was not included in the second questionnaire, instead an additional question on precise expertise was included

from collaboration within the Digital Earth consortium. This has clearly improved during the project, as initially only 32% of the respondents had found collaboration within the consortium after 11 months and underlines the importance of Digital Earth as a platform for inter-institutional cooperation. Considerably fewer respondents indicate they could have done the work alone (23 and 5% of responses in the first and second questionnaire, respectively). This may indicate that through collaboration between centres many new developments have been made possible. The wish for additional support in data science is also frequently mentioned, in about 26% of the responses in the second questionnaire.

We also identified how the Digital Earth project has contributed to progress on integrating data science methods in Earth System Science research, and for which goals (Fig. 8.1). Especially visual data exploration methods have contributed to data-gap closing, and improved scientific understanding within specific disciplines. AI approaches have contributed to data-gap closing, improved scientific understanding and to a lesser extent to data to proxy improvements. Other approaches such as other statistical methods and data quality assessment and controls have contributed to data to proxy approaches, and data quality and uncertainty specification, respectively.

As a very important requirement for data science, it was reported that appropriate observation instruments, data collection as well as data infrastructure are indispensable for doing data science. The Digital Earth project therefore would not have been possible without a strong basis of infrastructure and data. It builds on and profits from several complementary efforts focused on field observations and data infrastructure at the individual research centres, as well as targeted collaborative projects in this field.

With regard to science practices, several aspects of FAIR Science were reported. Most respondents make their data (77%) and Software code (67%) available, most

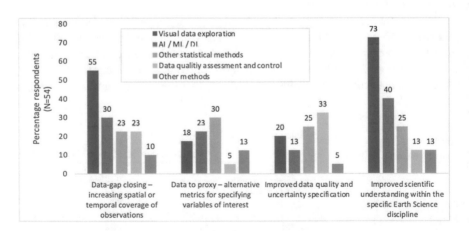

Fig. 8.1 Percentage responses to the question how different data science methods have contributed to achieving the overall scientific goals in the Digital Earth project

often through open-access repositories. However, licences and policies for the publication and use are not applied yet in several cases, and appropriate policy for licensing at the research organisations were reported as an important hurdle. Scientific and observational data, software tools and information are made available beyond academia, reported by about 50% of respondents in the final survey, which is a surprisingly large extent. This availability is complemented by the publication of guidelines for use, tailoring for specific applications and quality assessment.

The researchers report several important indicators that demonstrate the progress and scientific success of Digital Earth. First of all, researchers aimed to increase the usability of data, information and scientific workflows. In addition, they strived for better integration and collaboration between Earth/Environment—and data science disciplines. Joint scientific publications, conference presentations, new research proposals, and (open) software and data publications are regarded as most important signs of success for the Digital Earth project.

8.5   Conclusions

The evaluation reported here has been very useful in documenting the success of the Digital Earth project. The evaluation made use of criteria and indicators to assess the research capacities, goals and usability of results from the endeavour to adopt data science methods for Earth System Science. The framework and questions that are presented have made it possible to demonstrate and analyse the progress made during the project, as we have documented the capacities, goals and usability at the start and close to the end of the project, and the progress made during the collaboration between scientists from different disciplines. During the project, the criteria and indicators have been updated and extended, based on the feedback and evolving insights on what is required in terms of evaluation. In general, the criteria and sub-criteria and indicators presented here can also be applied to other projects in Earth System Sciences, but possibly also to other fields of research. We hope that other researchers and projects also feel encouraged to apply such an evaluation, in order to improve and progress in their research, and analyse and improve their success.

Acknowledgements We thank the Digital Earth project partners and project scientists for the good comments and suggestions for this evaluation during several project meetings, and for filling out the two questionnaires.

Appendix: Survey Questions

Category*	Included in first questionnaire	Included in second questionnaire	Question
C	1	1	Please indicate at which Helmholtz Center you are working
C	2	2	My role in Digital Earth is
C	3	3	My discipline is. Earth Scientist; Data scientist
C		4	What is your specific discipline?
C	4		How many persons do you have within your research team (team = department, or researchers) working on the Digital Earth project (DE)?
C	5		How many people in your team are working in the following fields of data science: AI / ML / DL; Advanced Data Visualisation; Data management
C	6		How are the capacities in your team for doing data science?
C		5	How are the capacities in your team for doing data science after the Digital Earth project? Have they improved?
C	7	6	With which data science experts from the following centres have you collaborated, within Digital Earth and externally?
C	8		With which Data Scientists and which data science institutes would you like to collaborate, within DE and external?
P	9		Where do you see the limitations of and challenges for data science?
P		7	In your opinion, on which of these limitations have we improved as a result of the Digital Earth project? And which ones do you still encounter?
P	10		What are the key observational and model (digital) datasets you will be using or producing?
P	11		Which observational infrastructure will you be using?
P	12		Which data and information infrastructure will you be using? This can be simulation infrastructure including HPC infrastructure, data storage infrastructure, etc
P	13		Which computer models (software/published code) or model output/data will you be using?
P	14		Will you develop new models or implement new model concepts?
P	15		What data science methods have you used already in other projects?
	16		What data science methods would you like to apply in DE?
P		8	What data science methods have you used in Digital Earth?
P	17		What is the purpose of applying these data science methods?

(continued)

(continued)

Cate-gory*	Included in first question-naire	Included in second question-naire	Question
P		9	Which scientific goals will you achieve/have you achieved? And which data science method(s) have helped to do this?
P	18		Which data exploration tools have you used already in other projects?
P	19		Which data exploration tools would you like to apply in DE?
P	20		What requirements do you have for data management?
P	21		Do you know best practices for data management, and are you using them?
P	22		Which scientific goals do you want to achieve by combining data, methods and data products, tools and knowledge from different disciplines (Earth Sciences, data science):
P	23	10	Which scientific goals of Digital Earth are most important to you?
P		11	What are the key products and tools you have produced? These include data, model (code), tools, workflows etc. Please provide a name/names, or brief description, or repository location
P		12	Have you been harmonising and integrating methods, models and/or data from different disciplines or research fields? If yes, did the Digital Earth project help to find a solution for this task?
P	24		Which research process goals are most important to you?
P		13	Which research process goals have been most important to you?
P	25		Which measures could be appropriate to assess the success of the overall goals of Digital Earth? Please specify the 4 most important:
P		14	Which joint output have you achieved through your work in Digital Earth?
U	26		Are your digital datasets (observations, raw data, model output) findable through a DOI identifier (see https://www.doi.org)?
U	27		Are you and your co-authors using the ORCID (Open Researcher and Contributor ID; see https://orcid.org) identifiers for scientific authors?
U	28		Are you using the IGSN (International Geo Sample Number; see http://www.igsn.org) identifier for geoscientific samples?
U	29	15	Are (parts of) your digital data (observational data, model output, and samples) accessible for other users?
U	30		Are there specific repositories you are using for storing data, accessible for other users?
U	31		Is there a metadata description for these data?
U	32		Will you develop specific DE computer code and scripts for data analysis? If yes, will this be accessible for other users?

(continued)

(continued)

Category*	Included in first questionnaire	Included in second questionnaire	Question
U	33		Are there technical standards (specific to your field of study) that you will apply for the data you will be developing?
U	34		Are your computer code and scripts using a formal language?
U	35		Are there Open Data and Software policies for the data, information and software tools you will be producing?
U		16	Has your computer code and scripts for modelling/analysis become available for other users?
U	36		What other principles are important? Please specify below:
U	37	17	Will results including data and software/scripts be made available for users beyond academia?
U	38		If yes, which of the following actions are taken? Usability assessment; User specific quality assessments; Tailoring of data products and methods; Guidelines for use and interpretation; Further support and services; Interface for data and knowledge of stakeholder actions; Other
P	39		Please give us any other comments, or ideas or suggestions for this questionnaire
P		18	What have been the five most important communication and information channels in Digital Earth for you? You can choose 5 at maximum
P		19	What has been the biggest obstacle for collaboration in Digital Earth? (for instance: transfer of information or data between centres)
P		20	What, in your opinion has been the biggest success of Digital Earth? (you can also give us any other comments)

*Categories are:
C = Capacities for data science
P = Project success and scientific progress
U = Usability of results

References

Junge K, Hadjivassiliou K, Kelleher J, Ramsden C (2007) Formative evaluation of the JISC VRE programme: the VRE1 programme: achievements and lessons learnt. The Tavistock Institute, p. 44. https://www.immagic.com/eLibrary/ARCHIVES/GENERAL/JISC_UK/J070900J.pdf

Stall S et al (2019) Make scientific data FAIR. Nat 570:27–29. https://doi.org/https://doi.org/10.1038/d41586-019-01720-7

Wilkinson MD et al (2016) The FAIR Guiding Principles for scientific data management and stewardship. Sci Data 3:160018. https://doi.org/https://doi.org/10.1038/sdata.2016.18

Wilkinson MD, Sansone SA, Schulte E, Doorn P, Bonino da Silva Santos LO, Dumontier M (2018) FAIR Metrics. https://doi.org/10.5281/zenodo.1065973

Chapter 9
Lessons Learned in the Digital Earth Project

Jens Greinert, Daniela Henkel, Doris Dransch, Laurens M. Bouwer,
Holger Brix, Peter Dietrich, Stephan Frickenhaus, Andreas Petzold,
Diana Rechid, Roland Ruhnke, and Wolfgang zu Castell

Abstract The Digital Earth project aimed for the integration of data science and Earth science. Here, we reflect on the main lessons learned that include the need for interdisciplinary collaboration, thinking out of the box, the concept of 'thinking in workflows' and models for the sustainable implementation of scientific software, data infrastructure and policies.

Keywords Digitalization · Earth System Science · Interdisciplinary · Collaboration · Workflows · Sustainability · Software · Data infrastructure · Policies

J. Greinert (✉) · D. Henkel
GEOMAR Helmholtz Centre for Occan Research Kiel, Kiel, Germany
e-mail: jgreinert@geomar.de

D. Dransch · W. zu Castell
Helmholtz Centre Potsdam-GFZ German Research Centre for Geosciences, Potsdam, Germany

L. M. Bouwer · D. Rechid
Climate Service Center Germany (GERICS), Helmholtz-Zentrum Hereon, Hamburg, Germany

H. Brix
Helmholtz-Zentrum Hereon, Geesthacht, Germany

P. Dietrich
Helmholtz Centre for Environmental Research-UFZ, Leipzig, Germany

S. Frickenhaus
Alfred Wegener Institute Helmholtz Centre for Polar and Marine Research, Bremerhaven, Germany

A. Petzold
Forschungszentrum Jülich GmbH, Jülich, Germany

R. Ruhnke
Karlsruhe Institute of Technology, Eggenstein-Leopoldshafen, Germany

W. zu Castell
Helmholtz Zentrum München, German Research Center for Environmental Health, Neuherberg, Germany

© The Author(s) 2022
L. M. Bouwer et al. (eds.), *Integrating Data Science and Earth Science*,
SpringerBriefs in Earth System Sciences,
https://doi.org/10.1007/978-3-030-99546-1_9

9.1   Introduction

A central question of the Digital Earth project is: How can data science contribute to improving scientific results within the Earth sciences? This fundamental question was posed by the scientists involved in the project, which has led to the research set-up and aims as described in Chap. 2. Within the project, several methods and tools have been developed and applied (see Chaps. 3, 4, 5, and 6). In addition, the collaboration and success of the project were assessed and evaluated (see Chaps. 7 and 8). In this chapter, we present as a conclusion the four lessons learned that we regard to be an essential basis for a fruitful interrelationship of data science and Earth System Science.

9.2   Lesson 1: Interdisciplinary Collaboration

Moving from multidisciplinary to interdisciplinary collaboration is essential for the adoption of data science methods and for making progress with digitalization in Earth System Science. One obvious success of Digital Earth is the established interdisciplinary collaboration. The results achieved in the project have been created by many scientists that before Digital Earth did not work together, did not know each other and might not even have seen the need and advantage of extending their own expertise before the project. Today, all project members agree that a sustainable collaboration across many disciplines, with different ways of working and despite the high number of obstacles and difficulties particularly in communicating with each other and finding a common ground, is a precondition for novel solutions and it is for sure worth the effort. A kick-start action, like the Digital Earth project, to such an endeavour is essential. Digital Earth provides the needed time to develop the collaboration and to establish a nucleus of knowledge and trust. Only this enables joint problem solving and develops new research ideas and opportunities.

9.3   Lesson 2: Thinking Out of the Box

Investing in working 'outside the box', beyond your own comfort zone, research centre and discipline is crucial. Communicating 'my' science to others and learning from other disciplines and approaches are important and the only way to expand our knowledge in Earth System Sciences. Cultural, language and importantly knowledge shortcomings hinder an effective communication and collaboration between data scientists and Earth scientists. This needs to be overcome by suitable means and strategies to help both sides acquire a good understanding of the other disciplines (see Chaps. 5 and 7). These shortcomings need to be done even if the process takes time and does not immediately lead to the envisioned success. Scientists with the explicit aim of bridging between disciplines, Earth compartments and institutions have been identified as good nuclei and multiplicators for developing and adapting novel data science approaches to improve Earth System Science. The Digital Earth project provided a frame to enable the distribution of knowledge within and across scientific disciplines and created an environment where people advanced beyond their typical realm.

9.4 Lesson 3: Thinking in Workflows

To set a common ground for interdisciplinary collaboration and 'thinking out of the box', the concept of scientific workflows was used in Digital Earth as a base for communication. After initial hesitation primarily by the Earth scientists, the concept helped to structure the processes of knowledge generation and to break it down into exchangeable and reusable steps. These workflows made it much easier to create a common ground between Earth and data scientists, to identify bottlenecks in specific steps in the workflow and to find alternatives for methods and tools. 'Thinking in workflows' (see Chap. 5) became the guiding principle in the project, where natural scientists define their needs, identify the available input data and present their wishes for output to the data and computer scientists. The computer scientists add their expertise with regard to methods and approaches in artificial intelligence, visualization, exploration of distributed data and software engineering. Thinking in workflows and formalizing the way Earth scientists generate knowledge allows an effective way of sharing and implementing scientific approaches and data science methods. It supports the reuse of scientific software and enables a component-based and collaborative framework for data-driven science. We identified the approach of 'Thinking in workflows' as a suitable and modular way of communication and scientific collaboration. Based on this approach, the next collaboration in smaller and larger projects will be much easier.

9.5 Lesson 4: Sustainable Implementation of Scientific Software, Data Infrastructure and Policies

The need for joined and professional software development and its maintenance is obvious when data science should become a cornerstone in Earth System Science. So far, such software is developed by small groups or individuals who train themselves. There is a need for more professional and standardized scientific software development. Software needs to be reusable and maintained to prevent scientists from inventing the wheel again and again. Research centres need to acknowledge that software development and maintenance is an ongoing and important effort similar to data management, running analytical facilities and the science itself.

Clear guidelines, policies and licensing rules for joint software development and provision, and the use of data are still 'under construction'. This creates problems when it is envisioned that developed software tools should be shared with others. More effort is required here.

There is no progress in data-driven science to be expected if infrastructure hurdles exist. Examples are data access (authentication) difficulties and the transfer of large data sets. This was experienced in the project and made collaboration and interdisciplinary research unnecessarily complicated.

Finally, the work and effort that are related to the sustainable implementation and development of scientific software, data infrastructure and policies have to be appreciated and counted as valuable scientific contributions.

Printed in the United States
by Baker & Taylor Publisher Services